T0269031

HUMAN RESILIENCE AGAINST FOOD INSECURITY

HUMAN RESILIENCE AGAINST FOOD INSECURITY

JOHN M. ASHLEY
Principal Food Security Adviser, Conseil Santé,
Clichy, France

ACADEMIC PRESS

An imprint of Elsevier

Academic Press is an imprint of Elsevier
125 London Wall, London EC2Y 5AS, United Kingdom
525 B Street, Suite 1650, San Diego, CA 92101, United States
50 Hampshire Street, 5th Floor, Cambridge, MA 02139, United States
The Boulevard, Langford Lane, Kidlington, Oxford OX5 1GB, United Kingdom

Notices
Knowledge and best practice in this field are constantly changing. As new research and experience
broaden our understanding, changes in research methods, professional practices, or medical
treatment may become necessary.

Practitioners and researchers must always rely on their own experience and knowledge in evaluating
and using any information, methods, compounds, or experiments described herein. In using such
information or methods they should be mindful of their own safety and the safety of others, including
parties for whom they have a professional responsibility.

To the fullest extent of the law, neither the Publisher nor the authors, contributors, or editors, assume
any liability for any injury and/or damage to persons or property as a matter of products liability,
negligence or otherwise, or from any use or operation of any methods, products, instructions, or
ideas contained in the material herein.

Library of Congress Cataloging-in-Publication Data
A catalog record for this book is available from the Library of Congress

British Library Cataloguing-in-Publication Data
A catalogue record for this book is available from the British Library

ISBN: 978-0-12-811052-2

For information on all Academic Press publications
visit our website at https://www.elsevier.com/books-and-journals

Working together
to grow libraries in
developing countries

www.elsevier.com • www.bookaid.org

Publisher: Candice Janco
Acquisition Editor: Louisa Hutchins
Editorial Project Manager: Emily Thomson
Production Project Manager: Omer Mukthar
Cover Designer: Vicky Pearson Esser

Typeset by SPi Global, India

What seest thou else
In the dark backward and abysm of time?

Act 1, Scene 2. Prospero to Miranda.
William Shakespeare, The Tempest.

CONTENTS

Acronyms   *xi*
About the Author   *xv*
Acknowledgments   *xvii*
Foreword   *xix*

1. Introduction   **1**

2. A Summary of "Food Security in the Developing World"   **7**
   2.1 Introduction   8
   2.2 Manifestations and Measurement of Food Insecurity   9
   2.3 Causes of Food Insecurity   9
   2.4 Mitigation of Current Food Insecurity   10
   2.5 Prevention of Future Food Insecurity   10
   2.6 Cross-Cutting Issues   11
   2.7 Conclusions   11
   References   12

3. Understanding Vulnerability to, and Resilience Against, Food Insecurity   **13**

4. The Anthropological Basis of Human Development   **21**
   4.1 General Introduction   21
   4.2 A Social and Cultural Obstacle Course   27
   4.3 Community Ownership   38
   4.4 Success Breeds Success   40
   4.5 Individual Food Security Strategies   40
   References   47

5. The Starting Point of a Development Intervention   **49**
   5.1 Introduction   49
   5.2 Personal Journeys to Our Understanding of "Food Insecurity"   58
   5.3 Seeking Consensus   61
   5.4 Challenging One's Assumptions   63
   5.5 Interaction With Local Administrations   66
   5.6 Lack of Trust Within Multiethnic National Communities   66

5.7 Conflict- or Political-Break With Tradition   67
5.8 Managing Expectations   68
References   69

**6.   Identifying and Prioritizing the Challenges Confronting
      Food Security Resilience for All   71**
6.1 The Need for Resilient Food Systems   72
6.2 Better Policy Making and Planning   76
6.3 Education to Build Resilience   91
6.4 The Peace Dividend   93
6.5 Other Priorities to Improve Resilience   104
References   109

7.   Building the Change Management Team and Approach   111
7.1 Advantages of Working Well as a Team   112
7.2 Getting It Right as a Team   115
7.3 Getting It Wrong as a Team   116
7.4 Institutional Perspective on Change Management   117
7.5 Program Implementation   119
Reference   123

8.   Importance of Local Knowledge in Building Resilience   125
8.1 Introduction   125
8.2 The "Groundnut Scheme" in Tanganyika   131
8.3 The Arctic Inuit   131
8.4 Sacred Sites in Liberia   133
8.5 Barking Dogs in Eritrea   134
8.6 Feedback on a Project in Central Asia   134
8.7 Crop Improvement Through Selection and Application   135
8.8 Harnessing Condensation for Drip Irrigation   137
8.9 Combining the Best of the Old and the New   138
References   144

9.   Lateral Thinking   147
9.1 Introduction   147
9.2 Flamingo Breeding   149
9.3 SWOT Analysis   150
9.4 Population Management   155
9.5 Bringing a New Idea to a Community   155
9.6 The Value of Corn Cobs in a Parched Field   161

9.7 Potato Promotion in France   161
9.8 Nepal Earthquake in 1998   163
References   165

10. The Role of Champions   **167**
10.1 Champions at Village and Public Sector Levels   167
10.2 Champions From the Commercial Private Sector   168
10.3 Champions Who Contest the Commercial Private Sector   170
10.4 Champions From International Organizations, Sport and Entertainment 171
References   172

11. Case Studies   **173**
11.1 Case Study 1. The Need for More Resilient Food Systems   173
11.2 Case Study 2. Resilience to Food and Nutrition Security Among the Inuit 190
11.3 Case Study 3. Human Capital as a Resilience Strategy in the Pamirs 202
References   207

12. Conclusions   **211**
12.1 One Person Can Make a Difference   211
12.2 Social Component Essential   211
12.3 Participation   212
12.4 External Actors in Development Design and Implementation   213
12.5 A Role In-Waiting for Social Anthropologists and Their Ilk   214
12.6 Building the Team Toward a Common Narrative   215
12.7 Sustainable Food System   216
12.8 Two Other Generic Priorities   217
Reference   218

Annex 1. Goats and Nightclubs of the Levant   *219*
Index   *231*

ACRONYMS

ADO	Agricultural Development Officer
AET	Agricultural Education and Training
AFMA	Agricultural and Food Marketing Association for Asia and the Pacific
AGIR	l'Alliance Globale pour l'Initiative Résilience Sahel
AKDN	Aga Khan Development Network
ASA	Association of Social Anthropologists (of United Kingdom)
ASEAN	Association of Southeast Asian Nations
ASTI	Agricultural Science and Technology Indicators
BBC	British Broadcasting Corporation
BRACED	Building Resilience and Adaptation to Climate Extremes and Disasters
CABI	Center for Agriculture and Biosciences International
CCC	UK Committee on Climate Change
CfP	Call for Proposals
CFS	Committee on World Food Security (of FAO)
CGIAR	Consultative Group on International Agricultural Research
CMAM	Community-based Management of Acute Malnutrition
DANIDA	Danish International Development Agency
DEFRA	Department for Environment, Food and Rural Affairs (United Kingdom)
DfID	Department for International Development
DPA	Displaced Person's Apathy
DRR	Disaster Risk Reduction
DVD	Digital Versatile Disc
EC	European Commission
ECHO	European Commission's Humanitarian Aid and Civil Protection department
ECOWAS	Economic Community of West African States
ERGS	Economic Recovery in the Gaza Strip
EU	European Union
FAO	Food and Agriculture Organization of the United Nations
FARA	Forum for Agricultural Research in Africa
GACGC	German Advisory Council on Global Change
GIZ	Deutsche Gesellschaft für Internationale Zusammenarbeit
GLASOD	Global Assessment of Human-induced Soil Degradation
GMO	Genetically Modified Organism
GSP	Global Soil Partnership
HIV/AIDS	Human Immunodeficiency Virus/Acquired Immune Deficiency Syndrome
IAASTD	International Assessment of Agricultural Knowledge, Science and Technology for Development
ICT	Information and Communications Technology
IDP	Internally Displaced Person
IDS	Institute of Development Studies, Sussex, United Kingdom
IFPRI	International Food Policy Research Institute
IMF	International Monetary Fund
INEF	Institute for Development and Peace (of Germany)

INGO	International Non-governmental Organization
IRA	Irish Republican Army
IRRI	International Rice Research Institute
ISAF	International Security Assistance Force
ITPS	Intergovernmental Technical Panel on Soils (of FAO)
LIFT	Livelihoods and Food Security Trust Fund (Myanmar)
LNGO	Local Non-governmental Organization
LRA	Lord's Resistance Army (Uganda)
LRRD	Linking Relief, Rehabilitation and Development
MBA	Master of Business Administration
MD	Managing Director
MDG	Millennium Development Goal
MDSI	Ministry of Development and Social Inclusion (of Peru)
MENA	Middle East and North Africa
MERAP	Middle East Regional Agricultural Program
MoBSE	Ministry of Basic and Secondary Education (of the Gambia)
MTE	Mid-term Evaluation
NCARE	National Center for Agricultural Research and Extension (of Jordan)
NEPAD	New Partnership for Africa's Development
NGO	Non-governmental Organization
NRA	National Resistance Army (of Uganda)
NRM	Natural Resources Management
OECD-DAC	Organisation for Economic Co-operation and Development's Development Assistance Committee
PDR	People's Democratic Republic (of Lao)
PDRY	People's Democratic Republic of Yemen
PIJ	*Palestine-Israel Journal*
PLO	Palestine Liberation Organization
PRA	Participatory Rural Appraisal
PRS	Poverty Reduction Strategy
PSNP	Productive Safety Net Program (of Ethiopia)
PTF	Petroleum Trust Fund (of Nigeria)
PVS	Participatory Variety Selection
RAF	Royal Air Force (of United Kingdom)
RE	Royal Engineers (of United Kingdom)
RRA	Rapid Rural Appraisal
SAARC	South Asian Association for Regional Cooperation
SADC	Southern African Development Community
SHARE	Supporting Horn of African Resilience
SOAS	London School of Oriental and African Studies
SOC	Soil Organic Carbon
SUN	Scaling-up Nutrition
SWOT	Strengths, Weaknesses, Opportunities and Threats
TA	Technical Assistance
TINP	Tamil Nadu Integrated Nutrition Project
ToRs	Terms of Reference
UN	United Nations
UNDP	United Nations Development Program

UNEP	United Nations Environment Program
UNESCO	United Nations Educational, Scientific and Cultural Organization
UNICEF	United Nations Children's Fund
UNRWA	United Nations Relief and Works Agency
VAHW	Village Animal Health Worker
VEW	Village Extension Worker
VIP	Very Important Person
WaSH	Water, Sanitation and Hygiene
WFP	World Food Program
WTO	World Trade Organization

ABOUT THE AUTHOR

John Ashley graduated in botany from London University, and then applied that knowledge to the field of agriculture for his doctorate from that university, working with the groundnut crop in Uganda. He also holds a degree in psychology from Cambridge.

Dr. Ashley has engaged in projects that have sought to help governments address current food insecurity, and increase resilience against future food insecurity. He has multisector program experience in agriculture and forestry, rural development, water, environment, education, nutrition and social transfers, roads and local government.

He has worked in some 30 vulnerable and/or conflict-prone countries for 40 years, especially in Africa and Asia. He was with the FAO for 5 years, and then became an adviser to national governments in interventions funded by international banks or donor agencies. He has conducted research with grain legume and cereal improvement programs in Libya, Kenya, Uganda and Nepal, and taught agronomy, crop physiology, ecology and human nutrition at Makerere University, Uganda.

Over some 45 consulting assignments since 1986, he has engaged across the project management cycle, including: formulating and designing development programs for the EU (economic growth in Somalia, and nutrition-sensitive social transfers in Nigeria), IFAD (Eritrea and Yemen) and DANIDA (Palestine), agribusiness investment proposals for the government/USAID in Liberia, and climate change response projects for Palestine; project implementation, such as a livelihoods project in Afghanistan (DfID), a regional agricultural research program in SADC countries (EU), two long-term projects in agricultural extension in Nepal (DfID, EU), and a nutrition program in Pakistan (EU); evaluations of the EU global Food Facility, wadi irrigation projects in Yemen, food security programs in Sudan and The Gambia, and economic recovery and job creation projects in the Gaza Strip; and capacity building with public and private sectors in north Africa and Jordan. Dr. Ashley's publications, as sole or contributing author, include four university textbooks on food and nutrition security, dryland farming and crop physiology.

ACKNOWLEDGMENTS

The author has given credits as necessary for generous permission from institutions and individuals to reproduce quotes. Every effort has been made to trace the copyright holders of publications cited, but if any have been inadvertently overlooked the publishers will be pleased to make the necessary arrangement at the first opportunity. The author also thanks those who offered comments on the textual material—Patricia Gadsby, Jessica Lumala, Dr. Perry Bosshart, Austin Ashley and Professor Philip Evans—and Martin Tayler for authorizing the inclusion of 10 of his photographs. Any errors or omissions throughout the book are of course the responsibility of the author. The views and opinions expressed should not be deemed as necessarily being shared by Conseil Santé of France, which health sector consulting company he advises in food security matters. Billie-Jean Fernandez, Nancy Maragioglio and Anita Vethakkan of Elsevier Food Science have been of enormous help in the book's production.

The author would like to offer thanks to all those who during his professional life have taught him so much of what he knows, from his early postgraduate days in Uganda in 1969 up to his current assignment in Pakistan—especially the farming, pastoralist and fishing communities whose company he has shared along the way. The learning curve was particularly steep in the early days, and he thanks those in university circles who guided him then in technical disciplines, in particular Professor Robert Whatley and Dr. Peter Moore of King's College London Botany School, Professor Bob Holliday and Dr. Bob Willey of Makerere University Faculty of Agriculture, Dr. Neil Fisher of the University of Nairobi, Professor Graham Milbourn of Wye College London, Professor Oliver Zangwill and Dr. Donald Laming of the Cambridge Experimental Psychology Department. Finally, thanks are due to Trinity College, University of Cambridge for providing the author with access to its facilities during the preparation of this book.

FOREWORD

This book is the follow-up sister volume to *Food Security in the Developing World* of 2016 by the same author and publisher. In that earlier book, "resilience against food and nutrition insecurity" received frequent mention, especially in Chapters 4–6. The term "resilience" also featured in the generic subtitle of the 2014 UNDP Human Development Report "Sustaining human progress: reducing vulnerabilities and building resilience", and means of resilience against food insecurity is being addressed as a major theme by the EC's assistance program. In June 2017 the EC published its latest Joint Communication on the subject, entitled "A strategic approach to resilience in the EU's external action" (European Commission, 2017).

The resilience theme is set against a special case of the "Tragedy of the Commons", an analytical albeit contentious essay published in *Science* magazine by Garrett Hardin in 1968. In the current scenario, depletion of communal environmental resources is occasioned not by selfishness or greed, but through increasing demands made upon the world's finite resources, of land and space, fresh water and food resources, by its exponentially burgeoning population. The latter is set to increase to 9 or 10 billion by 2050 from the current 7 billion.

According to the latest annual UN Report "State of food security and nutrition in the world 2017", published mid-September 2017, 815 million people were hungry during 2016 in terms of insufficient dietary energy intake. This number includes 155 million children aged under 5 who are stunted, and a further 52 million children suffering from wasting. The 815 million figure is 38 million higher than the number cited in the previous year's report, largely due to the proliferation of violent conflicts and climate-related shocks. The subtitle of the report makes mention of both resilience and peace (FAO, IFAD, UNICEF, WFP, WHO, 2017). Overlapping with those who are hungry, the WHO and other sources agree that at least 2 billion people, some 30% of the global population, suffer from micronutrient deficiencies, most of the victims of each category living in developing countries.

The focus on developing countries will continue in this follow-up book, though minority marginalized groups in developed countries will also be given their time in the spotlight, along with "transition" countries in Central Asia. Similarly, development rather than humanitarian issues will remain center-stage in the current book.

The tenor of this book, however, is rather different from the earlier one, which was factual and of a scientific bent. This follow-up book is more *raconteur*'ial in style, seeking to bring anecdotal evidence to bear on the subject of resilience from the current author's personal experience. Bear with me at my frequent mention of "I", a style which is new to me in my writings, yet which may offer variety from the equally ubiquitous "the current author"!

A suitable quotation to arrest the attention of the reader before engaging with this book is from Dylan Thomas' "A Child's Christmas in Wales", in which he lists the presents he received when a child, including "a book that told me everything about the wasp, except *why*". Those of us involved in fostering food and nutrition security of marginalized populations need to challenge ourselves each day not just on *what* we do, but *why*, *how* and *how best*, *where* and *with whom*. *What* is the easy bit, yet so often can be inappropriate because it has not been formulated in a well-informed and sensitive way. Often this is because local opinion has not been adequately sought and taken into account, including how or if the intervention can be absorbed by the target community or government. So often, only technical variables and change have been considered at the program planning phase, rather than equal address of the social aspects, such as capacity for behavioral change. It is imperative that planners understand that social change is a time-dependent *process*, whereby benefits are harvested by the community in perpetuity through behavioral change becoming socially-institutionalized. *Assumptions* on human behavior made by program planners are not always realistic.

Following an introduction in Chapter 1, Chapter 2 provides an overview of *Food Security in the Developing World* (hereafter "the earlier book"), while Chapter 3 presents a summary of what was said about vulnerability and resilience in that book. Chapter 4 considers why the human (anthropological) basis of decreasing vulnerability and building/securing resilience needs to be teased out in this follow-up book. The starting point of a development intervention is addressed in Chapter 5, while Chapter 6 examines how the challenges confronting the quest for food security for us all may be identified and prioritized. Building the change management team is considered in Chapter 7, and the importance of local knowledge in building resilience is the subject of Chapter 8. Chapters 9 and 10 consider the role that lateral thinking and "champions", respectively, can play in building resilience. Three case studies are presented in Chapter 11: the need for more resilient food systems; resilience to food and nutrition security among the

Inuit; and human capital as a resilience strategy in the Pamirs. Conclusions from the whole book are laid out in Chapter 12, and an Annex gives a roller-coaster stream-of-consciousness example of what can materialize as a result of lateral thinking.

REFERENCES

European Commission, 2017. A strategic approach to resilience in the EU's external action. Joint Communication to the European Parliament and the Council. JOIN (2017) 21 Final. June 7, 2017, 24 pp.

FAO, IFAD, UNICEF, WFP, WHO, 2017. The State of Food Security and Nutrition in the World 2017: Building Resilience for Peace and Food Security. FAO, Rome. 132 pp.

CHAPTER 1

Introduction

The need to improve food and nutrition security for the developing world is what economists and politicians may call a "motherhood statement", viz. "motherhood is good". As a concept it is anodyne, inoffensive and non-contentious. Yet to secure gains on the ground, a course needs to be charted through perilous terrain and waterways. This involves surmounting demand-side technical, economic, social, cultural and political hurdles at grass-roots level, and meaningful engagement with the supply-side vision, understanding and integrity of the people charged with delivering those gains in a sustainable way.

The original title conceived for this book was *The Anthropology of Resilience against Food Insecurity in the Developing World*. One of the reviewers of the author's initial proposal for this book rightly pointed out that what I had indicated I wanted to say went beyond the scope of anthropology, into allied fields of experience, such as sociology and psychology, so the title should be more generic. This advisory was accepted, and the title was adapted by the publisher to *Human Resilience against Food Insecurity*.

The current book refers to many learned publications of an anthropological nature that consider resilience in relation to food security. However, this will be done without the theoretical organization, analytical rigor and philosophical detachment for which some readers in academia may wish. Rather, as indicated already in the Foreword, a different approach has been deemed needful, one which the author believes better "connects" the non-specialist reader with the subject matter, as it is actually lived by the marginalized people concerned "in the field".

Such a rationale resonates with a comparable plea made for poetry in 2015, in an essay by California's Poet Laureate Dana Gioia (2015). He laments the demise of poetry, caused in large part by the college teaching of it being destructively analytical. "For thousands of years", he says, "poetry was taught badly, and consequently it was immensely popular. Readers loved the vast and variable medium of verse. It wasn't a forbidding category of high literary art; it was the most pleasurable way in which words could be put together". Gioia yearns for poetry to be made more accessible to the

Human Resilience Against Food Insecurity
https://doi.org/10.1016/B978-0-12-811052-2.00001-9
1

majority, to encourage more enchantment through this storytelling genre, and therefore more readers and writers of it. Such enlightenment can co-habit with the New Criticism, critical methods and literary theory as intellectual disciplines, yet not be pushed aside by them, destroying the actual holistic, intuitive experience of poetry.

"We need to augment methodology with magic", Gioia continues. "Poetry's existence on the pages of books, even the best-selling books, represented only a fraction of its cultural presence. Poetry flourished at the borders between print and oral culture—places where single poems could be read and then shared aloud". Referring to the current "disturbing trends" of teaching poetry, he says,

> When analytical instruction replaces the physicality, subjectivity, and emotionality of performance, most students fail to make a meaningful connection with poetry. So abstracted and intellectualized, poetry becomes disembodied into poetics—a noble subject but never a popular one. As the audience for poetry continues to contract, there will come a tipping point—perhaps it has already arrived—when the majority of adult readers are academic professionals or graduate students training for those professions. What is the future of an art when the majority of its audience must be paid to participate?

> *(Gioia, 2015)*

Likewise with food security, it should not be explained in terms merely of what, who and where, but also how and why, at the most basic level, that of human performance. The current author explores the anecdotal (the "oral culture") genre to convey the views of those at grass-roots level to the reader, rather than his writing just a turgid tome replete with specialist jargon which the generalist may find impenetrable. So, for example, instead of the term "degraded agro-pastoralists", a term from an anthropologist's lexicon, I talk of pastoralists who have settled. As other examples, I do not use the term "agency", beloved of sociologists and social anthropologists, referring to the actions individuals take to pursue goals (and their ability and willingness to do so), and I seldom use the term "paradigm", beloved of academics. Then every reader, not just the specific discipline specialists, will know what I'm about.

I much appreciate the travel writings of the Oxford-educated social anthropologist Nigel Barley, who is somewhat of an iconoclast in his field, and in so being has made anthropology accessible to non-anthropologists. In my earlier book I mentioned London's Museum of Mankind in its former home Burlington House, and how the 1976 exhibition I attended there, called "Nomad and City", blew my mind, with its mock-up of an Arab market place in Sana'a. Before long I was off to Yemen myself. Barley was a curator of that ethnological museum for some years.

When I use the term "anthropology" in this book it means what its Greek roots intend: "the study of humankind", as wide as that. Barley's book *Not a Hazardous Sport* is one of my favorite anthropological readings, though it is not on the student reading lists in many anthropological faculties.[1] In an interview with Rosie Milne in 2008, Barley said:

> As an anthropologist you're always asking questions such as: How different can different peoples be? Are we all reducible to a common humanity? And if so: what is it? Nobody can answer these questions. But I like to use fiction to try to answer anthropological questions. And fiction, I find, gives better answers.

> **(Daily Telegraph, 2008)**

While the current author will not employ *fiction* in the current book to help deliver such answers (though he will use *poetry* now and then), the purpose is to identify those answers (and the questions which occasion them) that may encourage a better "joined-up-ness" across the board. Just in case you missed it (!), the intent thereby is to empower better resolution of the many challenges that need to be addressed and overcome, to render vulnerable populations in the developing world more resilient against food and nutrition insecurity. For instance, the income of the millions of tea pickers worldwide is at best only just above the "living wage" level, making them highly vulnerable to the food access component of food security. It is clear that no one organization can improve their economic and social conditions; action is needed across the industry (Oxfam and Ethical Tea Partnership, 2013; Photo 1.1).

Both *causes* of food insecurity and *resilience* against it are amenable to mitigation by humankind. Human conflict is one such cause; wise leadership can promote its peaceful resolution. Disputes over land or grazing rights, say, are often an expression of population pressure, which is something that family planning can help address. Climatic extremes (droughts, floods, cyclones, etc.) are being caused/exacerbated by human activities, which can be modified by political will, long-term version, legislation and modified human behavior. And so on—such is the context of this book.

Bear with me when you note that the examples I give do not all have food security directly mentioned, in what some readers may find too discursive a style. Yet most actions we take each day are related directly or

---

[1] Just as Dana Gioia suspects he is generally *persona non grata* in university poetry departments http://danagioia.com/interviews/paradigms-lost-interview-with-gloria-brameg (accessed 5 July 2017).

Photo 1.1 Tea pickers in Darjeeling, West Bengal, India (April 2013). *(Courtesy of Martin Tayler.)*

indirectly to food security. Bringing resilience to communities may require lateral thinking, and wider consideration of *how* this may best be achieved. Annex 1 is particularly discursive and tongue-in-cheek, but it does culminate in the lady concerned adopting a food security strategy which rewards her with long-term resilience. Occasionally the reader might note a ray of humor to leaven the presentation, this being the author's survival strategy to get through the story of privation and hopelessness among the world's poorest people, whose own stories are seldom writ, other than in their faces and body language (Photos 1.2 and 1.3). It is hoped that this book might lead to a certain redress of their condition.

Photo 1.2 Slum life in Kolkata (originally Calcutta), capital of West Bengal State, India (April 2013). *(Courtesy of Martin Tayler.)*

Photo 1.3 Begging on the streets of Delhi (April 2013). *(Courtesy of Martin Tayler.)*

REFERENCES

Daily Telegraph, 2008. Fact and fiction. Interview of Nigel Barley with Gloria Brame. http://www.telegraph.co.uk/expat/4205428/Fact-and-fiction.html (accessed 5 July 2017).

Gioia, D., 2015. Poetry as enchantment. The Dark Horse poetry magazine. Summer issue. http://www.thedarkhorsemagazine.com/danagioiapoetrya.html (accessed 5 July 2017).

Oxfam and Ethical Tea Partnership, 2013. Understanding wage issues in the tea industry: report from a multi-stakeholder project. May, 32 pp. https://policy-practice.oxfam.org.uk/publications/understanding-wage-issues-in-the-tea-industry-287930 (accessed 6 November 2017).

CHAPTER 2

A Summary of "Food Security in the Developing World"

Contents

2.1	Introduction	8
2.2	Manifestations and Measurement of Food Insecurity	9
2.3	Causes of Food Insecurity	9
2.4	Mitigation of Current Food Insecurity	10
2.5	Prevention of Future Food Insecurity	10
2.6	Cross-Cutting Issues	11
2.7	Conclusions	11
	2.7.1 Companion Website	11
References		12

The forerunner to this current book was published in 2016, entitled "Food Security in the Developing World".[2] In it, the author provided what was intended to be a first primer on the subject for students and practitioners, covering its essential components, with an emphasis on "development" rather than "humanitarian" issues. The book was accompanied by a dedicated companion website, on which 10 case studies are provided which relate to generic issues of food and nutrition insecurity raised in the book itself, yet which were too lengthy to be included there.

The chapter comprises seven sections: (1) introduction; (2) manifestations and measurement of food insecurity; (3) causes of food insecurity; (4) mitigation of current food insecurity; (5) prevention of future food insecurity; (6) cross-cutting issues; and (7) conclusions. A summary of these sections is given as follows.

---

[2] Food and nutritional insecurity in the *developed* world was not addressed in the earlier book, the oft unclear distinction between the "developing" and "developed" world being discussed in its Preface, involving both economic and social indicators. In that book, the definition of "developing countries" was based on the UNDP Human Development Index, which addresses a composite of three indices measuring countries' achievements regarding citizens' life expectancy at birth, education and income.

Human Resilience Against Food Insecurity
https://doi.org/10.1016/B978-0-12-811052-2.00002-0

2.1 INTRODUCTION

An overview of the boundaries of the subject is presented. The context of the book was that in 2015, the Food and Agriculture Organization (FAO) estimated that just under 800 million people were undernourished in terms of dietary energy intake. Another manifestation of undernourishment—micronutrient deficiencies—affects about 2 billion people. Moreover, each year more than 3 million children die of undernutrition before their fifth birthday.

The term "food security" is often not well-understood by the general public, being equated with "food *availability*", for instance. Yet the term is far wider than that, including also economic, physical and social *access* to food, and the *nutritional, utilization, consumption* and *food safety* components of food security. Moreover, to assure food and nutrition security at homestead and community level, all these parameters need seamless integration, on a simultaneous and *stable* present continuous basis. The term "food security" is not merely about the *concepts* of hunger and undernutrition, but also the *experience* of these by those affected by or vulnerable to it.

The definition of food security adopted for the book was that agreed at the World Food Summit in November 1996: "Food security exists when all people, at all times, have physical, social and economic access to sufficient, safe and nutritious food which meets their dietary needs and food preferences for an active and healthy life". Conversely, the definition of food insecurity adopted in that book was that espoused by the FAO: "a situation that exists where people lack secure access for sufficient amounts of safe and nutritious food for normal growth and development and an active and healthy life".

As a whole, over the 20 years up to 2014, food supplies have grown more quickly than human population has in the developing world, resulting in rising food availability per person (FAO, 2014). This is in spite of global food supplies exhibiting larger-than-normal variability in recent years, reflecting the increased frequency of extreme climatic events such as droughts and floods. Though improvements in economic access to food are indicated by reduction in "poverty rate" (which fell from 47% to 24% between 1990 and 2008 in the developing world as a whole), economic access to food (based on food prices and people's purchasing power) has fluctuated more recently. The vital aspect of constrained "access" of one sort or another to food was underlined by Amartya Sen in his statement that "Starvation is the characteristic of some people not having enough food to eat. It is not the characteristic of there being not enough food to eat" (Sen, 1981).

The hunger, wasting, stunting, and underweight manifestations and measurement of food insecurity are broached, together with the interlinkage between food insecurity and nutrition insecurity, and the downstream effects on individual and community health. Worldwide distribution, numbers, and condition of the food-insecure are discussed, as well as parameters of short- and long-term food insecurity. Food as a rights issue and food and nutrition insecurity as a gender justice issue are considered. The concept of investment as a potential apex solution to food insecurity is tabled, together with a summary history of feeding the world since 1945, and the UN organizations most involved with food and nutrition security issues.

2.2 MANIFESTATIONS AND MEASUREMENT OF FOOD INSECURITY

The concepts of hunger, undernutrition, and food and nutrition security are expanded from their presentation in Chapter 1. The five key elements of the UN Secretary General's "Zero Hunger Challenge" of 2012 are listed, based on a conviction that hunger can be eliminated in our lifetimes. Nutrient deficiency diseases are explored in more detail, both protein-calorie and micronutrient (minerals and vitamins), as well as the increasing prevalence of nutrient oversufficiency, leading to overweight and obesity conditions, the majority of the global increase since 1980 being in developing countries. The challenges of measuring household food security and undernutrition are discussed.

2.3 CAUSES OF FOOD INSECURITY

The multidimensional technical and social etiology of food and nutrition insecurity is considered in detail, the potential causes often reinforcing each other. These include: poverty and insufficient awareness; environmental degradation and climate change; food price hikes and price instability; conflicts; weak institutional environment (uncoordinated policies and institutions, and insufficient investment in agriculture and public services); and predisposition to diseases, and intestinal affliction. Large-scale land lease by one country in another, and large-scale arable land set aside for biofuel production are two potentially contributory factors that are also considered.

2.4 MITIGATION OF CURRENT FOOD INSECURITY

This chapter discusses how the food insecurity being experienced by individuals and communities is being addressed. Coping strategies at microlevel, humanitarian aid, and Food-for-Work and Cash-for-Work schemes are considered. The Euro1billion EU Food Facility instrument launched in response to the 2008 food price crisis is reviewed in some detail, together with examples of other donor programs to enhance agricultural productivity and marketing. EC Communication 586 of 2012 "Enhancing the Response to Crises" is highlighted. Four examples of interventions specifically redressing nutrition insecurity are given: (1) the decentralized Community-based Management of Acute Malnutrition (CMAM) model, and (by contrast) long-term chronic malnutrition addressed through focus on maternal nutrition during pregnancy and lactation, and protecting the health and nutrition of the child during the first two years of life, through community-based approaches covering health, diet, and care; (2) interventions to address micronutrient deficiencies—food fortification and dietary supplements; (3) The Copenhagen Consensus and the Scaling-up Nutrition (SUN) Movement; and (4) government-owned social transfers to promote nutrition security in northern Nigeria.

2.5 PREVENTION OF FUTURE FOOD INSECURITY

Eight parameters are considered relating to how food insecurity threats to the most vulnerable may best be addressed. These comprise: (1) anticipating crises by assessing and mitigating risks and vulnerability, through (a) early warning systems, (b) focus on prevention and preparedness, and (c) decreasing vulnerability and building resilience; (2) national food security strategies; (3) increasing food availability (a) through exploiting natural resources, such as "Green Revolution" techniques, modified to render them more environmentally-friendly; Participatory Variety Selection (PVS); physiological and biochemical research; biological control of crop and forest pests, and human/livestock disease vectors; improved feeding of ruminant livestock; support to artisanal fisheries, and (b) reducing food loss and wastage; (4) addressing better access to, and utilization of, nutritious food, through (a) employment and job creation programs, and (b) public policies for social protection; (5) addressing both availability and access together, through (a) strategic food reserves, and (b) better identification of investment and trade opportunities; (6) addressing nutrition security through biofortification; (7) safeguarding food safety; and (8) enhancing the stability aspect of food security.

2.6 CROSS-CUTTING ISSUES

Nine cross-cutting issues are discussed, comprising: (1) the right to food; (2) gendering food security; (3) land tenure; (4) population growth pressure on land and water resources; (5) empowerment and resilience through access to factors of production; (6) food security governance; (7) government capacity to formulate, implement, and institutionalize change and reform; (8) food sovereignty; and (9) climate change.

2.7 CONCLUSIONS

Six main conclusions are advanced. In terms of removing the causes of food and nutrition insecurity, top priority was accorded to enhancing women's empowerment and gender transformation.

2.7.1 Companion Website

The 10 case studies on the companion website comprise: two from Africa (the Productive Safety Net Program in Ethiopia, and "Conservation Agriculture" in Southern Africa); three from Latin America and the Caribbean (causes and remedies for undernutrition and poverty in Belize, food security challenges from natural and man-made disasters in Ecuador, and regional food and other insecurities in Amazonia and La Plata Basin); and two from Asia (causes of food insecurity in the mountainous north of the People's Democratic Republic of Lao (Lao PDR), and among the Kuchi pastoralists of Afghanistan). The remaining three case studies were of a global nature, addressing the achievement or otherwise of Millennium Development Goal (MDG)1c relating to hunger and nutrition, the development and effects of the food price crises of 2007/2009 and 2011, and urban slums as epicenters of deprivation and food insecurity. The latter scenario was depicted as the main picture on the front cover of the book itself, showing an aerial view of *Favela da Rochinha*, the largest slum in Brazil on the forested hillside of Rio de Janeiro, with the smart high-rises of the former capital city in the background.

The book and website combined, however, can give only an overview of the complexities of the multifaceted subject, and further reading within the array of specialist fields (technical, economic, anthropological, sociological, cultural) will repay attention for the advanced student of rural development, not to mention the literature on the humanitarian assistance aspect of food/nutrition insecurity. Moreover, the connection between the two, linking relief, rehabilitation and development (LRRD), is a vital area for the student to interrogate, with its many worthy publications of bilateral development agencies, international organizations, and nongovernmental organizations (NGOs). These between them cover the conceptual approach, funding, prioritization, and practical on-the-ground issues, the integration and coordination of which is often suboptimal.

The issue of building resilience of individuals and communities against vulnerability to food and nutrition insecurity is receiving increased attention of late, with the idealized intention of reducing the need for unsustainable humanitarian assistance. In my 2016 book "Food Security in the Developing World", there were many references to "vulnerability" and "resilience", especially in Sections 1.7 and 5.1 and Chapters 4 and 6, with 145 citations altogether. These issues were therefore given high priority, indeed in each chapter. Furthermore, on the book's companion website, all 10 case studies addressed resilience and/or vulnerability; there were 59 citations altogether. A summary of these vulnerability and resilience references is given in Chapter 3 of this book, to create the platform for the additional treatment of each provided in this follow-up book, as they are the overarching theme therein.

REFERENCES

FAO, 2014. In: The state of food security in the world 2013: an overview. Asia and Pacific Commission on Agricultural Statistics, 25th Session, Agenda Item 9, Vientiane, Lao PDR, February 18-21. FAO Statistics Division. 5 pp.
Sen, A.K., 1981. Poverty and Famines: An Essay on Entitlement and Deprivation. Clarendon Press, Oxford. 257 pp.

CHAPTER 3

Understanding Vulnerability to, and Resilience Against, Food Insecurity

Resilience against food and nutrition insecurity as discussed in the earlier book is summarized in this chapter. The concept of "resilience against food and nutrition insecurity" has two dimensions—not merely the inherent strength of an individual, household, community or larger entity to better resist stress and shock, but the capacity for it to "bounce back" and recover relatively rapidly from the impact. "Resilience" in that book, and this one, is defined as "The ability of an individual, household, community, country or region to withstand, adapt and quickly recover from stresses and shocks". *Food security* is but one element of *livelihood security*, and indicators of the former should not be interpreted independently of a good understanding of the latter. This is discussed further in the "about this companion website" introduction to the case studies on the earlier book's companion website. The FAO's more elaborate definition of "resilience of people and their livelihoods" incorporates this concern, namely:

> *The ability to prevent disasters and crises as well as to anticipate, absorb, accommodate or recover from them in a timely, efficient and sustainable manner. This includes protecting, restoring and improving livelihoods systems in the face of threats that impact agriculture, nutrition, food security and food safety.[3]*

The earlier book cites the 2014 UNDP Human Development Report "Sustaining Progress: Reducing Vulnerabilities and Building Resilience", and discusses the interconnected two concepts and their importance in securing human development. The UNDP Report shows that although overall global trends are positive, lives are being lost, and livelihoods and food and nutrition security undermined by continued natural disasters and man–made crises.

---

[3]http://www.fao.org/emergencies/how-we-work/resilience/en/.

Human Resilience Against Food Insecurity
https://doi.org/10.1016/B978-0-12-811052-2.00003-2

Though having the potential to save lives and relieve suffering, humanitarian aid is unsustainable. Better by far if such aid can be merged with development assistance, structured to build resilience in the community.

Section 5.8 of the earlier book addresses the difficulty of assuring the *stability* component of food supply. Regions of the developing world that are prone to drought and/or cyclones and flooding, the economies of which are still modest, are the most vulnerable to food insecurity. Support is needed for such vulnerable populations because even a transitory crisis can trigger chronic food problems, as assets are quickly depleted and livelihoods undermined. Increased volatility in agricultural markets, leading to a sudden price rise in food and/or fuel, has underlined the acute vulnerability of the poorest to even a small additional shock, for example, a poor harvest (due to drought or a plague of field pests like locusts or weaver birds).

Even in a good harvest year, there will be a fluctuation in domestic food supply of the farming community, relating to the "hungry season", which occurs before the staple crop in the field is ready for harvest yet the homestead granary is empty or almost so. Food security stability is conditioned by geography too. West Asia, North Africa and the Caribbean are particularly vulnerable because of their limited or frail natural resource base, and their heavy reliance on international food markets for domestic supplies.

Food insecurity is particularly hard to tackle in complex ongoing crises and the tenuous transition to stability. During a crisis, fragile states may lack the capacity or institutional framework to implement long-term food security solutions, a situation often compounded by poor governance, conflicts, man-made disasters, malaria, measles, HIV/AIDS, Ebola and other diseases. Where states' governments lack the capacity or the will to address the risks and needs faced by the most vulnerable people, international resources continue to play an important role. Section 2.5 of the earlier book discusses differential vulnerability to food and nutrition insecurity, at community and individual levels. The communities most at risk include the rural poor, urban communities in conflict and slum dwellers, while at individual level it is children, women, the handicapped, sick and infirm.

The FAO has established that agricultural production needs to increase by 60% between 2005 and 2050 to enable the projected population of 10 billion people by then to be fed. Already, increasing population pressure on land and water is reducing agricultural productivity and environmental integrity. The earlier book points out that climate change will likely intensify that pressure in the developing world, adding to the difficulty of reconciling food needs with production potential. Food production, storage, distribution and use systems are needed that are equitable and sustainable in all dimensions: economic, social and environmental. On the production side, a more resilient food system can in part be brought about through better technical efficiencies of irrigation systems or introduction of more drought- and/or saline-tolerant crop varieties.

Climate change will affect all four dimensions of food security: food availability, food accessibility, food utilization and food systems stability. It will have an impact on human health, livelihood assets, food production and distribution channels, as well as changing purchasing power and market flows. Its impacts will be both *short term*, resulting from more frequent and more intense extreme weather events, and *long term*, caused by changing temperatures and precipitation patterns. *People who are already vulnerable and food-insecure are likely to be the first affected.* Climate change modeling suggests that the tropics will be most affected by climate change. Many of the poorest countries are there and are likely the least able to adapt; hence they require the most help to do so. Agriculture-based livelihood systems that are already vulnerable to food insecurity face immediate risk of increased frequency and depth of crop failure, new patterns of pests and diseases, lack of appropriate seeds and planting material, and loss of livestock. People living on the coasts and floodplains and in mountains, drylands and the Arctic are most at risk. Dryland agriculture in arid and semiarid regions, where over 40% of the world's population live, including more than 650 million of its poorest and most food-insecure, is particularly vulnerable to drought and increased risk of food insecurity for its residents. Also, in many coastal regions there are large agricultural areas which are at risk of increased salinity of ground water and flooding, associated with rise in sea level; this is exacerbated in cyclone-prone regions.

One of the important recommendations in the High Level Panel of Experts (HLPE) (2012) report (in its Section 5.1) is that climate adaptation policies and programs should be complementary to, not independent of, those fostering food security. Climate change is but one of a range of threats to food security. Interventions designed to increase the resilience of food systems are also very likely to contribute to climate change adaptation. Farmers should be at the center of these efforts, and location-specific approaches need to be devised ensuring that community needs are met, based on local knowledge, skills and other assets.

Section 1.7 of the earlier book pleads for greater investment in climate-resilient *smallholder* agriculture, currently responsible for producing most of the world's food, and improving the enabling environment in terms of policies, and access to markets and finance. From a long-term perspective, support to sustainable agriculture is paramount to build up resilience in sub-Saharan Africa, for example, where the sector provides employment to 60% of the population, including the most vulnerable.

Chapter 3 of that book considered the multiple causes of food security, with Section 3.3 addressing *environmental degradation*. A paragon of participatory planning excellence was cited as the Productive Safety Net Program (PSNP) in Ethiopia, embedded in government structures and budget, in which LRRD principles have been followed. Through Cash-for-Work public labor, the environment has been improved through re-sculpting the hillsides with terraces, water breaks and reservoirs. This renders the environment less prone to run-off and surface erosion, and more capable of absorbing and retaining the water it receives, so it can be stored in the soil profile or as water sources for livestock. The social transfer of money for public works enables food to be purchased from local markets, which will themselves become better supplied by produce from the local watershed. This is a self-reinforcing strategy of intervention which confers resilience while providing a ready exit strategy for the PSNP trust fund donors.

A second cause of food insecurity, cited in Section 3.6, is the *weak institutional environment* which often pertains in affected countries, in which there are uncoordinated policies and institutions. Donors' initiatives to address this include the EC's "l'Alliance Globale pour l'Initiative Résilience Sahel" (AGIR) partnership, launched in 2012, which provides a resilience roadmap, which reinforces existing West African regional strategies. The AGIR aims to build the resilience of vulnerable populations by consolidating responses to food crises and structural causes of food insecurity, as well as building resilience to chronic undernutrition.

Chapter 4 considers the mitigation of *current* food insecurity, one means being microlevel coping strategies (Section 4.1). Negative coping strategies, which *fail* to cope, need to be rendered unnecessary through building resilience. Elsewhere in Chapter 4, the resilience-building Euro1billion EU Food Facility instrument and some other donor interventions were explained, along with EC Communication 586 of 2012 offering insights into ways to improve the impact of donor partner responses to crises through an enhanced resilience approach.

Chapter 5 of the earlier book addresses prevention of *future* food insecurity. Section 5.1 concerns anticipating crises by assessing and mitigating risks and vulnerability, and the need for a focus on prevention and preparedness. Early warning systems are vital, as is formulating national food security strategies to encourage donor partner buy-in. Removing the causes of vulnerability, such as poverty, instils resilience. The global proportion of people living on less than $1.25 per day fell from 43.1% in 1990 to 22.2% in 2008. However, the food, fuel and financial crises since 2008 have worsened the situation of vulnerable populations and impacted poverty reduction programs in some countries.

Recent and recurrent food crises in the Sahel region and the Horn of Africa have underscored the need to address chronic vulnerability and develop a long-term and systematic approach to building the resilience of vulnerable countries and populations. Subsistence farmers, pastoralists and slum dwellers must feature strongly in the target groups. Enhancing resilience lies at the interface of humanitarian and development assistance, and calls for a long-term approach and complementarity, ensuring that short-term actions lay the groundwork for medium- and long-term interventions. The approach should be based on alleviating the underlying causes of crises and reducing the intensity of the latter's impact, through enhancing local/national/regional coping and adaptation mechanisms and capacities to better manage future shocks, uncertainty and change.

It was suggested in Section 5.1.3 that resilience strategies of nations and their development partners, such as the EU, should contribute to a range of policies, in particular regarding Food Security, Climate Change Adaptation and Disaster Risk Reduction (DRR). As pointed out by the EC in its COM 586 document (Section 4.6), resilience can be successfully constructed only in a "bottom-up" manner. Partner countries, their policies and priorities, must take the lead. The food price riots of 2008 have propelled governments of many countries to reform policies and change practices in support of national preparedness and building resilience.

Action to strengthen resilience should be based on sound methodologies for risk and vulnerability assessments, the latter serving as the basis for elaborating and implementing national resilience strategies and disaster reduction management plans, as well as for designing specific projects and programs. A vulnerability analysis is needed before any intervention is planned let alone implemented, to optimize project design.

In response to the massive recent food crises in Africa, two of the initiatives which the EC is pursuing are AGIR, as mentioned above, and "Supporting Horn of African Resilience" (SHARE), which set out a new approach to strengthening the resilience of vulnerable populations through learning from experience, in order to multiply and scale up successful approaches and actions. The SHARE and AGIR initiatives represent an improvement in the way humanitarian and development forms of assistance interact, boosting levels of assistance in the short-term, facilitating the link between relief, rehabilitation and development, and demonstrating the commitment of the EU to address the root causes of food insecurity in the longer term. The two initiatives focus on food security in sub-Saharan Africa, though this approach can equally be applied to other regions and other types of vulnerability (e.g., regions threatened by floods, cyclones, earthquakes, droughts, storm surges and tsunamis, climate change or food price increase).

Investing in resilience, as advocated above, is cost-effective. Addressing the root *causes* of recurrent crises is not only better than responding only to their *consequences*, but is also much cheaper. All development partners are coming under increased pressure to deliver maximum impact for the funds available. In the cost-effectiveness context, the PSNP in Ethiopia, one of the largest social transfer schemes in sub-Saharan Africa, which provides for transfers in the form of food or cash to the most vulnerable households in the country in return for participation in public works, costs only about one-third of an equivalent humanitarian intervention.

Generically, social protection measures to ensure nutritious food for the most vulnerable include strengthening the resilience of their communities and livelihoods to shocks from conflicts, climate change and natural disasters. There have always been individuals who have fallen through the kinship safety net, incurring loss of all forms of security and dignity, perhaps adequate shelter and companionship, and vulnerability to hunger, undernutrition and poor health, even to the point of death. Protection of these individuals could enable them to stay above the threshold of mining their assets beyond that critical level in order to survive crises so that it becomes impossible to recover, like selling the thatch on the family house to buy food.

Internationally-funded food security programs need to reflect partner government and regional policies and processes, not only to address current challenges but also to anticipate future shocks and build resilience against them. Yet such efforts at building resilience do not have to rely solely on donor partners. A heart-warming community effort in Tigray province of Ethiopia is a case in point (BBC News, 2015).

As observed by the Committee on World Food Security (CFS) in the First Draft Principles, there is significant evidence that investing in agriculture and food systems is one of the most effective ways of reducing hunger and poverty, through building stronger and more resilient communities

(FAO, 2014). Unfortunately, the very regions where food insecurity and poverty are most widespread are the regions where agricultural investment has been stagnant or declining.

Chapter 6 of the previous book introduces the subject of needing to *engender* food security initiatives. As readers of this book will know, women have many vital traditional roles in building resilience against food insecurity crises in the developing world, at household and community levels. In rural areas, home to the majority of the world's hungry, as well as taking care of the home and looking after children, it is the women who grow most of the crops for domestic consumption and are primarily responsible for all postharvest activities, including processing and preparing food. They also handle livestock, gather fodder for stall-feeding and fuelwood, and manage the domestic water supply. Moreover, they undertake a range of community-level activities that support agricultural development, such as soil and water conservation. Yet women's work often goes unrecognized by their menfolk at home.

In addition to discrimination at home, it is rife at community level too, where women lack the leverage necessary to gain access to resources, training and finance. According to a study by the World Bank (2012), 103 of the 141 countries studied have laws and cultural barriers that can thwart women's economic aspirations through limiting their access to key resources.

These obstacles not only heighten their vulnerability to personal food insecurity but also reduce their contribution to family agricultural production. Furthermore, it should be recognized that women are especially vulnerable to natural disasters. For example, in Myanmar during Cyclone Nargis in 2008, 61% of those killed were women.

Tackling a community's social and economic access to food and good nutrition requires a more politically-engaged approach, which challenges head-on the gender dimensions of poverty, addressing gender–inequitable power relations and norms, both inside the household and beyond. Social protection programs such as cash and voucher, and food-for-work schemes may be empowering for women in some respects, but are often not gender *transformative*, instead often reinforcing the existing *status quo* of unequal gender roles and relations. The program Building Resilience and Adaptation to Climate Extremes and Disasters (BRACED) is helping build resilience, especially of women to disasters, with the Department for International Development (DfID) providing grants to 21 projects led by NGOs to scale up activities in South Asia and Africa with a focus on improving agricultural practices. Other forms of support may also be required for farming, geared toward satisfaction of family and community food needs, through serving the values of livelihood resilience, autonomy and stability rather than being oriented toward profit and market competiveness.

Furthermore, the Institute of Development Studies points out that the gender equality potential of short-term food security solutions is being limited by narrow targeting of groups perceived as the most vulnerable, with an almost exclusive focus on pregnant women and very young children (IDS, 2015). As a result, the needs of other vulnerable female groups, such as older women, adolescent girls, women from marginalized communities, single women and women with disabilities, are rendered invisible.

REFERENCES

BBC News, 2015. In: Haslam, C. (Ed.), Turning Ethiopia's Desert Green. April 20. http://www.bbc.co.uk/news/magazine-32348749.

FAO, 2014. CFS Principles for Responsible Investment in Agriculture and Food Systems First Draft (for Negotiation). 13 pp.

HLPE, 2012. Food security and climate change. A report by the high level panel of experts on food security and nutrition of the committee on world food security, Rome. HLPE Report 3, June. http://www.fao.org/3/a-me421e.pdf.

IDS, 2015. Gender and food security. In: Brief. Bridge Development—Gender. Institute of Development Studies, Sussex. 8 pp.

World Bank, 2012. Women Business and the Law 2012: Removing Barriers to Economic Inclusion–Measuring Gender Parity in 141 Economies, January 1. 167 pp.

CHAPTER 4

The Anthropological Basis of Human Development

Contents

4.1	General Introduction	21
4.2	A Social and Cultural Obstacle Course	27
	4.2.1 Clearing the Hurdle	28
	4.2.2 Falling at the Hurdle	32
4.3	Community Ownership	38
4.4	Success Breeds Success	40
4.5	Individual Food Security Strategies	40
	4.5.1 Sudanese Goatherd	41
	4.5.2 Sudanese Shoe Shiner	42
	4.5.3 Fishing Community	43
	4.5.4 Tomato Growing in the Sudanese Desert	43
	4.5.5 Sale of Fodder in Sudan	44
	4.5.6 Police in Sudan	45
	4.5.7 Police on the Owen Falls Dam, Uganda	45
	4.5.8 Police on the Fort Portal Road, Uganda	45
	4.5.9 Room Cleaner in The Gambia	46
	4.5.10 Private Enterprise in Refugee Camps	46
References		47

4.1 GENERAL INTRODUCTION

Peggy Bartlett (1980) sets out to delve below generalized macro-issue discussions on food production, transfer of technology and economic development to consider individual farming families, their microlevel decision making and the choices they make about their farm operations, their use of resources and their long-term strategies for survival. Their decisions are affected by global and regional processes, which in turn determine the products each nation has to export and the *resiliency* of the country concerned in a food crisis.

Though it is almost 40 years old, some of the above volume's contributory chapters by economic anthropologists (and associated social science "brands") are just as topical today, the whole being fairly free of dense

Human Resilience Against Food Insecurity
https://doi.org/10.1016/B978-0-12-811052-2.00004-4

anthropological jargon. The studies presented demonstrate the practical contributions that the social sciences can make to agricultural development programs and policy. Better understanding of the agricultural decision-making processes should lead to better government policies and strategies, and wiser, more-informed interventions funded by international donors.

Bartlett's book contributes to a better understanding by outsiders of the complex web of factors that influences real-life decision making of smallholders and hunter-gatherers in developing countries, and thereby of the process, challenges and potential of rural development. These decisions are against the backdrop of an ever-starker Anthropocene era—faced by climate change, environmental decay through overexploitation and pollution, increasing inequality and poverty, growth of urban slums, still-dreadful undernutrition and health indicators, to name but a few, *and* "our" global need to feed an extra 2 billion people on the planet by 2050 (mostly in the developing world).

The social environment for decision-making by a given farmer includes decisions made by others. Decisions that may seem irrational to an outsider, conforming to tradition or expending resources on "social investments", are clearly rational when set in their social context (Berry, 1980). The sociological concept of each person's somewhat plastic "identity" plays a part too, whereby the micro-social processes of our interaction with society help shape our identities—how they are created, maintained, challenged and reinvented. Conformity to social norms, face-saving behavior and institutional identities are among the topics covered in a book by Susie Scott (2015).

As expressed by Cohen (1967):

The principle of economic rationality ... is, in fact, a combination of three assumptions ... The third assumption states that, given enough information, men (that is, humankind) ... will seek to maximize their gains by obtaining the highest possible return for any given resources, or that they will seek to use the smallest quantity of resources to obtain a given return.

Chibnik (1980) notes that anthropologists have long had concerns about the applicability of economic rationality (one of the cornerstones of classical economic theory) in the context of the rural developing world, especially in regard to the multiplicity of goals of the decision makers therein. The current author would also add that a maximizing strategy is often too risky for smallholder farmers.

Allen Johnson (1980) in his chapter in Bartlett confronts the inadequacy of mathematical optimization studies to explain the economic decision making of farmers in developing countries. The aforesaid methodology, though bringing intellectual rigor to "the field" with the purpose of prescribing behavior, is insufficiently flexible, particularly in its ability to anticipate and cope with creativity of human solutions to economic problems. He furthermore notes that agricultural decision making is full of surprising behavior (to the outsider), which does not necessarily prioritize rigorous optimization[4]. Johnson speaks of "... farmers who have lived in an environment their entire lives, observed countless details about its soils, crops, weather, labor supply, market prices and government intervention, and have integrated these experiences with cultural 'rules of thumb' into a total understanding that all our research methods in combination can hardly fathom". To better understand heuristic[5] "agricultural decision making", he enthuses about the potential to provide a better insight by an ethnographer "who develops the ability to apprehend the whole context of decision making by living in the locale long enough to overcome ethnocentric bias and sense the rationality by which decisions are made locally".

In the same volume edited by Bartlett (1980), Gladwin and Murtaugh, in their exploration of cognitive processes, propose that decision making occurs in two modes—the attentive and pre-attentive—which helps to explain how past decisions are integrated into behavior and choice patterns. The authors caution agronomists and social scientists to be aware that "pre-attentive" knowledge rather than attentive thought governs many decisions and actions which farmers make. Finally, Cancian (1980) points out the fallacy of believing that people generally act on knowledge, that they use this knowledge to calculate, and having calculated, act. Often people are called upon to act before they can "know".

Development economist Jean Drèze (2017) also challenges the belief of mainstream economic theory that "rational self-interest" is the prime motivation of humanity. The book cited is a volume of essays based on his research and field surveys from 2000 to 2017. He maintains that people act from love, compassion, public-spiritedness and so on, and that this is the route to liberty and equality. Public policy needs to be far better informed by everyday experience of the poor—solutions to their poverty and hunger must go beyond government "projects". Drèze aspires to what India

---

[4]Among pastoral people though, optimization of grazing in relation to rainy and winter seasons is clearer to see (Case Study 7.1 of the earlier book companion website).

[5]An heuristic technique is any approach to problem solving, learning or discovery that employs a practical method not guaranteed to be optimal or perfect, but which is sufficient for immediate goals.

would look like if policies and politics were truly driven by public interest. Government policy there actually consolidates the divide between the haves and have-nots in his view. He sees a change in the making though, with street-level activists challenging the power of politicians, big business and the corporate-sponsored media.

That other economists also have reservations about traditional "economic rationality" theory has been recognized by the award of the 2017 Nobel Prize for Economics to Richard Thaler, a founding father of the school of "behavioral economics", and its associated "nudge theory". Humans are not the rational beings as portrayed by classical economic rationality theory, Thaler argues. He has explored how psychology and an individual's environment shape economic decisions, often leading to bad or irrational decisions, rather than optimizing, due to insufficient thinking time, habit and straightforward poor decision making. Often, people opt for the easiest decision rather than the wisest. Supermarkets have long been "nudging" shoppers to buy items additional to the "essentials" for which they visited the store. Researchers have demonstrated that putting healthy food at eye level on supermarket shelves increases sales, by "nudging" their decision making, even when the shopper has scant knowledge about nutrition or the cause of obesity. Thaler was co-author with Cass Sunstein of the best-seller book "Nudge: Improving Decisions about Health, Wealth, and Happiness" in 2008. This influenced both governments and large businesses, leading, for example, to the United Kingdom government in 2010 starting a "nudge unit" (more officially called the "Behavioral Insights Team"). This team, together with its international partners, is applying behavioral approaches to challenging and diverse issues, such as their work in refugee camps in Jordan and Lebanon, and seeking to persuade people to undertake better long-term planning, for example by starting a pension.

I learned so much about Kyrgyzstan, its varied peoples, and its literature from my national colleague Zura, when on October 31, 2015 we took time out from our project identification work and climbed Mount Suleiman together. There we looked over the city of her birth, Osh, which straddles the old Silk Route across Asia. I tested with her my observations on behavior of people in Osh's famous fruit and vegetable market (mainly women, as traders and buyers), such as the apparent lowly status in the pecking order of the few Russians who stayed on in the country after the Soviet collapse in 1991. At other markets close to Jalal-Abad I observed that the marketing and processing of walnuts was also dominated by women. An abiding memory of Osh market for me was the passion displayed by the pigeon fanciers, all of them male, who quietly occupied a small space on the edge of that market showing off and trading their variously-traited birds. Without an intimate understanding of a country and its spirit, it is surely more difficult to fathom the most impactful interventions and how best they may be seamlessly woven into the fabric of a society.

In neighboring Kazakhstan, it is important for an outsider to understand how much rural Kazakhs treasure their horses. There are more than 20 Kazakh words for "horse", each with its own shade of meaning, and more than 30 words or expressions to describe the colors of a horse's coat. This linguistic richness reflects the importance of the horse, as part of "cultural identity", just as there are many words for "snow" or "seal" in Inuit culture, for instance (Section 11.2.2.1). The majority in the "West" have become uncoupled from rural life since the era of agricultural mechanization began with its associated shedding of rural livelihoods, and have lost that oneness with nature and the concept of the "sacred balance" (see Section 8.1). This puts them "on the back foot" when going into the developing world, yet how many stop to ponder on that? An understanding of ruralness and its importance to national identity in developing countries, even for those who over recent decades have joined the urban drift there, has to be re-learned or learned from scratch, if the "Westerner" is to understand the context of what he or she is trying to achieve to improve resilience.

On an island in the Bay of Bengal off mainland Rakhine State, Myanmar, which I visited in 2017, my suggestion that the desperately scarce drinking water supply could be increased at homestead level, by collecting rainwater from the tin roofs of houses, encountered the cultural imperative that rain first has to strike the ground before it can be used. Another example of a cultural constraint on water use I came across in Garanda, Borno State, northeastern Nigeria in 1991. I observed water gushing from a borehole, forming a huge pool at which livestock were drinking. I queried the sustainability of this, and was told that engineers had put a regulatory cap on the borehole, yet the pastoral community had knocked it off, on the grounds that water is God-given and should not be restricted. In 1999 I returned to that Sahelian zone site—the water was still gushing, but at a far lower pressure. Before too long that aquifer will be unproductive. The dire situation as described in both Borno and Rakhine surely implies a role for social anthropologists to try to open a debate with local communities and their leaders to improve their resilience against water shortage, and food and nutrition insecurity downstream of that.

Those who seek to grapple with how villagers and others in the developing world make decisions are confronted with the state of flux and ferment that characterizes the developing world today. Societies are dynamic, under the influence of massive information inflows and different ways of thinking from many exogenous cultures.

To personalize this, my (Ugandan-born) wife and I have observed in Uganda the huge change in culture of the Baganda in Central Region (formerly called Buganda), especially since the mid-1980s. The reasons include a sequence of dictators having been thrown out, the influence of returnees from political exile in developed countries, and the outreach effect of satellite television channels, the internet and social media into even far-flung rural parts. As a result, life has become better-informed, and a whole lot faster and more urgent. We can be sitting in an airport lounge anywhere in the world and, if we are minded so to do, hear on WhatsApp family members telling us of the latest unreasonable and risible behavior of our neighbors in the village where we farm (behavior usually occasioned by the poverty in which they live, and their struggle to make ends meet). In an instant, we can send money by cell phone to one of our staff in the village to pay his child's school fees, when he finds himself unable to do so himself. It seems like yesterday, when 50 years ago I first landed in Uganda to work at Makerere University, and hardly needed to look each way before walking across Kampala road, as there was hardly any traffic; now the streets are jammed with vehicles, including the deadly *boda boda* motorcycle taxis that seem oblivious to the rules of the road.

Another example of rapid change, from southern Africa this time, is given by Susan Kent (1996a), who cites the rapid cultural evolution undergone by the Basarwa she had been studying in Botswana over just 6 years from the late 1980s, albeit that these were recently settled people who formerly had a dispersed nomadic lifestyle. "Change has been so rapid and so encompassing that, had I not witnessed it myself, I would have been skeptical of someone else's description of it", she says. She rightly points out that the high rate of change in these modern times will likely not have happened in the past. She uses the term "degraded agro-pastoralists" to refer to recently settled Basarwa people in the Kalahari. Greater severity of drought has driven people to settle, knowing that the government provides emergency rations to settlements now and then, in times of crises. Their change in lifestyle was a rational decision.

Kent (1996b) explains that goat herds were not viewed as sources of meat by the Basarwa in their former sustenance hunter-forager lifestyle, but as potential emergency rations should hunting be unsuccessful—they provided a buffer against the insecurities of hunting and foraging. In the past, mobility was the buffer against poor hunting. With sedentism, mobility is no longer an option and goats kept in "opportunistic herding" mode provide better food security for the family—the goats wander off on their own in the day, the owner hoping they return at night, otherwise a search party will be sent out the next day. Less care is given them than a pastoralist would give. Kent also discusses networks of food sharing, another strategy of food security among the Basarwa.

Section 4.2 comprises an anecdotal survey of the types of challenges and "mis-cues" that can occur at the start of a cooperative venture as partners from very different backgrounds endeavor to find common ground and engage with each other. It is better to grapple with such "occupational hazards" upfront, as it may reduce the incidence of pained hand-wringing later, at our failure as individuals or a group to judge adequately our approach and performance to bring respite and resilience to those whose lives, livelihoods and securities are at risk.

4.2 A SOCIAL AND CULTURAL OBSTACLE COURSE

Any development intervention in a developing country community can proceed only as quickly as is anthropologically possible. Chapter 1 of this book opened with the observation that on the way to improve individual or community resilience against food and nutrition insecurity, there are many "hurdles" which some of the implementers involved will likely face, including those of a social and cultural nature. These need to be negotiated successfully to facilitate a seamless transition from "before-project" through "project implementation" to "post-project" sustainability. An outsider, even a fellow national from another part of the country, will likely be unsighted on the norms that condition the engagement of the "beneficiary community" in an intervention that could lead to incremental income generation and/or food production, for example. These hurdles relate to variant perspectives, expectations and attitudes, and different starting points.

It may take some time for the various starting-line perspectives of development partners to "gel", to find common ground and empathy, before real progress can be made on development work (which in the context of this book is building resilience against food and nutrition insecurity). There is a minefield of cultural accidents waiting to happen "out there". It is so easy to cause or take offense unwittingly, while so many potential mistakes and misinterpretations can be foreseen and avoided with a little preparation, coaching and mentoring by those with an inside knowledge of the culture and social norms concerned. Unfortunately, the logic that we should all try to understand the intent of the speaker or doer is often foregone in the immediacy of our reaction. Several generic examples of clearing, or falling at, these initial hurdles are given below.

4.2.1 Clearing the Hurdle

If a good rapport is established upfront, the resulting friendship lays in "credits" for future work together as equal partners.

4.2.1.1 Ankole District, Uganda

At the start of my life in East Africa as a postgraduate student, I paid the school fees of a boy I met in Kampala. As a result of this I was taken to his village right on the equator in a pastoral part of southwestern Uganda called Ankole, to meet his family. Following protocol I was first introduced to the traditional chief of the area in his thatched mud house. As my eyes became accustomed to the relative darkness in the hut, illuminated only by two small unglazed windows, the shutters of which had been opened, I could see the chief sitting on a chair, with a semicircle of men cross-legged on the floor before him. Intuitively, I joined the group of seated men, and we were all served local yoghurt (which has a strong smoky aroma and taste)[6]. Though I did not understand the vernacular conversation that followed, my schoolboy friend later told me that the chief had said he had never before seen such love expressed by an outsider for his community.[7]

4.2.1.2 Return of the King

In 1969, the Kabaka (King) of Buganda[8], Sir Edward Mutesa II, died in exile in London, to where he had fled from his palace in 1966.[9] Two years after his death, in 1971 when political conditions in Uganda allowed, his body was brought back to Buganda in a British Royal Air Force (RAF) VC 10 aircraft. Before the plane landed at Entebbe, the plane flew low and slowly

---

[6]Milk can be kept "fresh" for a long time under such remote conditions, if it is stirred with freshly charred sticks, which imparts a smoky taste to the milk or derived yoghurt or butter.
[7]Later still this schoolboy (Labby Tugumisizi) was clubbed to death by the rifle butt of a guard of the district capital town administrator to whom Labby had gone to query a decision that the administrator had made.
[8]A south-central region of Uganda, in which most of the Baganda ethnicity reside.
[9]An event brilliantly recounted in Mutesa's autobiography "Desecration of my Kingdom".

around Buganda so his people might participate in his final journey. It flew right over me as I worked in my experimental plots at the university farm. At each wingtip was an RAF escort fighter, diminutive in scale compared with the VC10, which was the state-of-the-art plane at the time, regarded by many as one of the most beautiful aircraft ever built. It is one of those indelible moments in life, something impossible ever to forget. I explained to my largely-Baganda team of field workers what they were witnessing. I felt that the enormity of the moment did not reach them at that time, whereas I was deeply moved. Only later when the news coverage caught up with the event did the Baganda reluctantly accept that their king had died and his body come home, and acknowledge the friendship, empathy and goodwill accorded them by the former Protectorate power. The British and Ugandan peoples (all ethnic groups) continue to have cordial relations, though there are intermittent spats at government-to-government level.

4.2.1.3 Upper Barclay, Eastern Nepal

From 1987 to 1990, I was an agricultural adviser in a rural development program funded by DfID. During this assignment I once visited a village, Upper Barclay, with Sankhuwasabha District's Agricultural Development Officer (ADO) L.N. Galal and a couple of his staff. The village is at about 2,500 m, in the foothills of Mount Makalu, one of Nepal's 8,000+ m peaks. We lodged with a family in its traditional two-story wooden house. I recall the wild bush which served as the local source of tea, the goats being inflicted with blood-sucking leeches, and the huge docile buffalo. These provided animal draft for both rainfed (*bari*) terraces, and the *khet* valley terraces, where monsoonal rice was grown followed by post-monsoon wheat, using water stored in the soil. In addition, the multitasking buffalo threshed the grain from the harvested crops by walking over it, carried jerry cans of water from the valley to the homestead, provided milk for the family, much of which turned into yoghurt (*dahi*, दही), and manure for the fields. Moreover, and not least, the huge-framed beasts provided rudimentary central heating during the cold winter months as they are kept on the ground floor of the house, below the human living quarters, their body heat rising through the slatted wooden floor.

To provide entertainment for the family, I danced with the ADO on the living room floor to the sound of taped Nepalese music. When all were tired, the family and visitors all slept together around the central hearth, radiating out from it like a huge sunflower, men and women, old and young alike. In this way, we visitors entered into family life on their terms, while

being able to provide some advice which would help improve their livelihoods. It was a partnership of equals, with reciprocal benefits in abundance.
The ADO masterminded the whole exercise, and a year later proceeded
to Edinburgh University on a Master's course in agricultural extension at
British government expense, which I had arranged for him. I had learned
the basic "dos" and "don'ts" of Hindu and Buddhist cultures while I was
in Kathmandu, before taking up residence in the eastern Kozi Zone of the
country. If there had been other cultural "hurdles" for me in this traditional
mountain home, I was led over them by the masterful ADO, without touch-
ing them or even noticing they were there.

4.2.1.4 Meeting the Ancestors

In 2011, I was formulating some agribusiness investment proposals in Nimba
County, Liberia, on its northern border with Guinea. One of these targeted
the defunct cocoa industry, which was in the doldrums, largely related to
the massive internal displacement of the farming community during recent
civil strife. While I was inspecting some of the first cocoa trees brought into
the country by an early settler from Guinea, planted as an understory in
natural hardwood forests, one huge tree was pointed out. I enquired, "Is this
tree sacred, where you meet the ancestors?". "Yes". My showing this awareness and empathizing with it clearly impressed the cocoa farmer. He and I
were able to sit and discuss our different experiences of this phenomenon,
his in Liberia and mine in western Uganda. That white people, whom I
represented, could find such commonality was scarce believable to my new
friend. The gulf between the two colors seems rather wide in rural Liberia.

4.2.1.5 Motorbike Accident, Liberia

In a remote village, I came across a young mother of three, Bendu, who was
slowly dying, having fallen off a motorcycle taxi, smashing her leg. Bacterial
infection of the fractured bone had set in, and her leg was about to go gan-
grenous, yet her family could not afford any medical treatment costs. Her
community had tied a cotton thread around her ankle in the belief that this
alone would bring about healing. To cut a long story short, I took her to
Monrovia, where a Rwandese orthopedic surgeon first treated the osteo-
myelitis with antibiotics, and then performed the necessary operation, saving her life. I visited her each evening while she recovered. Other patients
on the ward were split between those who thought I was God, and those
who thought I must have been the one who knocked her off the motorbike.
Having a white person as a hospital visitor to an African each evening was

difficult for them to comprehend. How little it took to break down that cultural barrier, with some friendship and sensitivity. Stereotypes cannot withstand hands-on experience. How Bendu cried when my contract was over and I had to leave.

4.2.1.6 Concern at Colleagues' Wellbeing in Afghanistan

In February 2010 the Taliban attacked two guest houses in Kabul, which were commonly used by visiting development workers. I had myself stayed at one of them, Park Residence, on an early mission to that country, but wasn't in the country at the time of this attack. Apparently Indian nationals were being targeted, to express Taliban distaste for Indian bilateral support to Afghanistan. Nine Indian guests were killed. "My" guest house closed and relocated with its staff to another site. On my next mission to Kabul I sought out this new location to enquire if any of the staff whom I knew had been hurt in the attack. I was surprised how this concern touched the staff, that a foreigner should have gone out of his way to enquire about their wellbeing. Such goodwill generated, for so little effort, the commonality of humanity shining bright in a troubled area beset with powerful players still pursuing their Great Game.[10]

4.2.1.7 Creating Space

Space has to be created in which an expatriate's counterpart can hold forth. All too often, the expatriate comes seeking only to know how the national can conform with the expatriate's norms and needs. Listening is an art form, a process. It takes time. On a recent program evaluation I undertook I was surprised by the response of one of my counterparts, the head of an international organization in the country concerned which was implementing part of the program; she herself was a national of a developing country. In a private email to me she said:

> I wanted to take the opportunity to really thank you for your work. I mean, the approach you have taken as an external evaluator is a "modern one", the one that all evaluators should take. That is, besides reviewing documents and meeting different people, and visiting the field, to discuss project issues before, during and after; to openly analyze findings with different stakeholders; to be very open to listen

---

[10]I witnessed a similar instance when I was in Kuwait in November 2013, when Christine Lagarde, Managing Director of the IMF, during a visit took time out to sympathize with the Philippine expatriate community, whose home country had just been devastated by Typhoon Yolanda (Haiyan).

*to different opinions and to even investigate and look for additional information
and adjust findings and recommendations if deemed necessary; to discuss good
achievements and encourage stakeholders; but also to discuss failures to achieve
targets, and ask stakeholders to identify solutions! Very useful indeed!*

My "surprise" was my disappointment that this experienced professional
expressed surprise, indicating that my team's approach to the evaluation was
"unusual".

I was greatly impressed by the humility expressed by René Devische
(1993) in the prologue of his ethnography on the Yaka in Zaire, in the
face of his craft on the one side, and the knowledge that he will never
completely know the people he lived with for some 3 years. He quotes
with approval Dan Rose (1990) who favors an ethnography that "… urges
a moral and aesthetic practice: do radical ethnography, one that gets you
closer to those you study at the risk of going native and never returning; it
is hoped, at least, that you will not again embrace the received assumptions
with which you (inheriting your academic texts, methods, and corporate
academic culture) began".[11] These are sound guidelines for any foreigner,
ethnographer or development worker, going into the field.

4.2.2 Falling at the Hurdle
4.2.2.1 Tripoli, Libya
At the other end of the scale of success is failure. In 1980 when I worked
in Libya on an agricultural development project with the FAO, one of my
colleagues, a Sudanese who had taken his Ph.D. from the same London college that I had, said he would visit me one day at my home in the United
Kingdom. "Good", I had replied. "Do let me know when you are coming".
My response caused him great offense. "So you can arrange to be out!"
he retorted. "No", I replied calmly, "so I can arrange to be in!" Yet serious
damage had been done to our friendship. In *his* culture, my comment was
impertinent, though not in *mine*. I will not make that particular mistake
again. I will say only, "It will be nice to see you". If ever he comes and I am
elsewhere, which is most likely, we will both be sorry.[12]

---

[11]Nigel Barley was mentioned in Chapter 1 as a somewhat iconoclastic social anthropologist,
and Dan Rose is another.

[12]Thirty-six years later, when I worked in Sudan, I tried to find him, but he was out of country—I could not tell him I was coming (!), as I found out his whereabouts through another
Sudanese colleague only when I had reached Khartoum.

4.2.2.2 A Tanzanian's Sensibility

A similar instance of mis-communication was told me by my Tanzanian Kiswahili instructor, of how offended he was when he visited an English friend in upcountry England. He was met at the railway station by his "friend", who after the initial greeting asked him when he was planning to return to London. The intention of his host was good, to be helpful, so he could enquire with the stationmaster on train times enabling him to bring his Tanzanian visitor back to the station just in time for his train, so he would not be inconvenienced by a long wait. Yet the Tanzanian interpreted the enquiry about his return time as meaning that he was not a welcome visitor, and that his host couldn't wait to get rid of him, even before he had been taken to his home.

4.2.2.3 Sensitive Language and Customs Related to the African Colonial Period

During my postgraduate days at Makerere University I made friends with an undergraduate Kenyan in my faculty. He kindly invited me to his home in the uplands above Nakuru, anxious to show me the large "fields" of crops near his home, so different from the smallholder agriculture plots around Kampala, and I spent a lovely weekend with him and his family. Back in Kampala, there came a time when he ran short of money to pay one of his young brother's school fees in Kenya, and we came to an arrangement that I would lend him the money. As I recall, he delayed in getting the money together to repay me and apologized. I found myself saying to him, "Oh don't be silly, Reuben. That's quite all right". Bang!. I'd used the word "silly", which had caused him huge offense. It rang the colonial "bell" for a Kenyan, which rang loudly. I labored to explain to him that colloquially in England, "silly" is an endearing term. It does not equate to "stupid" as he had interpreted it. I told him that when my father had called me "silly" I would not at all have taken offense. Yet I was never able to talk my way out of that hole I'd dug for myself.

Interestingly, shortly after the incident when I was on home leave in United Kingdom, I had absent-mindedly mused to my father whether a Morris Mini tire would fit into my briefcase. "Don't be silly", he said. "It will not". I found myself bristling slightly, as Reuben had changed my own perspective on the word "silly", such that it could be regarded as offensive!

In later years when I worked in Kenya, I found out the position Reuben held in the Kenyan government, and wondered whether I should try again to explain. Yet I decided against it, as my earlier attempts at reconciliation had been so painful for us both. Our friendship had collapsed through my careless use of a word I had not intended to mean anything other than an expression of friendship.

I later observed that another "colonial" word—"boy"—could cause offense. My humble colleague Dr. Dan Davidson on a Ugandan development project had referred to my young counterpart James as a "nice boy" to his girlfriend. She gave Dan a serious lecture on how he must never use that term again in that context. The word was used by the British in the colonial period in East Africa to denote their house "servants"—houseboy (or housegirl). I felt so sorry for Dan, as he had not intended to suggest that James was of a lowlier status than himself, but quite the opposite, to compliment him. Dan, a white American, was completely unsighted on any colonial hang-ups in East Africa. Moreover, Uganda was not administered under the British mandate as a colony as Kenya was, but as a protectorate, the British and Baganda serving each other's interests, so I was myself surprised at this response by the young Muganda lady. In the protectorate days, the British made the Baganda more "colonial" than the Baganda had themselves been before the British came, by sending them to other areas of the country to administer it, and giving them "privileges". Interestingly, well-to-do Ugandans in Kampala, including my relatives, still refer to their house staff as "houseboys" and "housegirls", this being quite acceptable within that urban grouping, black to black.

During that period of working in (dryland eastern) Kenya to which I referred above, I was able to come to the rescue of a British colleague who had offended the local police in Machakos township. There is a rule that each day when the national flag outside of the town administration office is being raised in the morning and lowered in the evening, passing pedestrians and vehicles must stop. Unaware of this government protocol, my friend had ridden his bicycle past the flag-lowering ceremony, and was arrested. I happened to come along shortly afterwards and witnessed him arguing his case with the officers, on the grounds of his ignorance. This was not impressing the police, as ignorance of the law is no defense when one transgresses it. Had he apologized, he would have been sent on his way with a warning, but that had not been his approach. I intervened in the dispute, telling my friend he had done wrong and apologizing to the police officers on his behalf. Only then was he free to continue his evening. It was probably the colonial regime which had instigated the protocol, and a British citizen who had infringed it. It is a respectful custom, and I'm sure my friend had learned from his experience to be more culturally-aware in future.

As a final example, my late friend Hugo Barlow, the Ugandan Makerere University farm manager, told me that he used to get upset when he was young and would walk into a British administration office and hear from a white British administrator, "What can I do for you?". To Hugo, this relegated him to an underclass in his own country, and was still disquieting for him decades later! (A general practitioner doctor I once had in Cambridge greeted me once in his surgery not with "What can I do for you?" but "What are we doing today?"—this was an amusing, user-friendly, and hugely endearing greeting. He was saying to me that "we both, doctor and patient, will work together as a team in resolving the matter, whatever it is". Well done Dr. Marshman!)

4.2.2.4 Sensitivities to Being "Beneficiaries"

Some communities can be particularly sensitive, for example, about being made to feel recipients of largesse. This can be avoided with the appropriate body language and adept choice of words. It is after all in the rich world's self-interest that global food and nutrition insecurity is eradicated, so it is incumbent upon the rich world to imagine itself in the position of today's poor (see Section 10.4). In Nepal, many an expatriate has got into trouble by suggesting to Nepalese "beneficiaries" that they are "doing something *for* them", and that the Nepalese should cooperate. I've witnessed great offense being caused through this manner of presentation. I recall an incident in which a dear Nepalese friend of mine took exception to such a presentation being made to a meeting of villagers by my Scottish colleague on the development team. I felt so sorry that both of my friends had been hurt, when no offense had been intended by the Scotsman.

A way needs to be found whereby the outside specialist comes into the society as an equal cooperator in a joint venture, and is prepared to move at the speed with which the community there is prepared to move. Once this formula is established, real progress can be made and great friendships forged. I know this from having worked in that country for 5 years. I particularly recall witnessing the frail mother of one of my Nepalese staff in Arghakhanchi district becoming extremely distressed with worry that something could happen to me in 1998, as Maoist sympathizers started to infiltrate its main town Sandhikharka. She felt sure that I, as the only foreigner living in the district, would be killed,[13] as the insurgents, who

---

[13]Shortly afterwards, one of the project vehicles was set alight during the night as a warning, and the donor agency instructed me to leave the district immediately. Several years later in 2002, a large guerrilla force attacked the local police station at night, killing many of the police inside and burning government installations.

wished to topple the government, viewed any development project as assisting the government rather than the people.

Just as the Nepalese people justifiably feel very proud of their long ancestry and are sure that they are "something special" among God's creation, I later found a similar national pride while working in Myanmar. Despite being there for only a month, I became convinced that many of the foreigners working there in "high places" failed to appreciate that Myanmar nationals were very proud of their heritage too, strongly resenting being told by outsiders that they were thinking and doing things in the wrong way. They long to be understood on their terms, not through the perspective of modern imported moral and political criteria suited to those foreign nations. The Burmese are justifiably aggrieved that many foreign agency staff seem not to trouble to understand them, what *they* the nationals think about events in their country and why, and how those thoughts are rooted in history. Nor do they feel that foreigners in high places in Myanmar get to grips with the political tightrope that the Chief Counsellor has to walk between what she may wish to do and what the military authorities require to happen for their "security" concerns, as the military and Burmese people view most of the countries which border Myanmar as hostile and exploitative. I met many intellectually-impressive and lovely Burmese people when I was in that country, and learned so much from them about a part of Asia with which I was previously unfamiliar.

"Space" for such debate may best come about in an informal setting. A fascinating learning experience for me was courtesy of my Burmese colleagues, who felt able to share with me their vision on what was needed to bring peace to their country, and the behavioral changes within-country which were needed for that. The most important of these latter seems to be that the majority Burmese Buddhists need to learn how to accept "The Other", namely the Muslim minority in Myanmar. I also learned from them how such internal change was hampered by foreign donors and experts who could not escape from their own pre-formed perceptions of the country context and politics, finding it difficult to adjust their theories to Myanmar reality. Far from preventing the tragic death and displacement of Muslims in Rakhine State, this ignorance by "Western outsiders" involved in "development" activities serves only to exacerbate the internal conflict, thereby reducing resilience against food insecurity of both Buddhist and Muslim communities. Most telling were the Burmese accounts of foreign donor staff and consultants having little interest in even listening to national opinion, regarding local consultants as nothing but vassals to ensure that foreign consultants had their meetings arranged, and arrived for them, in time. Something must be seriously wrong on the Western outsider's part surely, when so many educated Burmese are angry at them, and vowing never to comply with recommendations in some of their reports? This "space" is one that cannot be closed on the current trajectory (see Section 6.4).

It is not only expatriates who can get into trouble in situations where those doing the intervening have not properly gelled with the community and explained themselves. In 2016 in Sudan, I heard a story in the remote Aulieb zone in Red Sea State. A past project had donated a pump to a community to lift water from a well, and national project staff visited to monitor progress at site, and take photos of the pump in action. Before long the visiting staff were confronted by an irate villager: "We know you think we have sold the pump. That is why you keep coming back to make sure you are right! So take your pump back! We don't want it any more!".

4.2.2.5 Kyrgyzstan: How to Create a Diplomatic Incident Without Trying

In Kyrgyzstan in January 2016, a Scotsman, employed as a welding superintendent at a mining company posted what he considered a joke on Facebook, concerning the favored Kyrgyz national dish of *chukhuk*.[14] His aim was to raise a few cheap laughs from his mates, but the repercussions almost led to a "diplomatic incident", as a result of nationals tuning in to the conversation. The posting led to vitriolic responses from many Kyrgyzs, including his work colleagues. He was arrested by the police and informed that what he did could put Kyrgyzstan on a war footing with the United Kingdom. He faced racially-motivated hate charges that could have landed him in jail for 5 years. He was fortunate that the authorities merely deported him and imposed a ban on his returning to the country for 5 years.

One Kyrgyz blogger posted on Facebook that the strong adverse reaction in the country to the Scotsman's comment is a symptom of something deep in her society. She maintains that there is a well of discontent towards anyone who is not Kyrgyz by blood, a result of the nation's anger towards strangers who could be spies, and that this can so easily erupt into physical violence.

This Kyrgyz story goes to show that it really is worthwhile for a foreigner to take time out to understand "The Other" country and culture in which he or she works and lives, not just in terms of spectacular mountain

[14]http://www.rferl.org/content/kyrgyzstan-kumtor-horses-penis-outrage-british-expat-mcfeat/27468972.html.

scenery and handicraft, but "what makes people tick", what grievances they may have about being exploited, and so on, from both a psychological and a cultural perspective. The story is no different from that of a Kyrgyz visitor to Scotland having to dig a lot deeper than the glens and grouse moors there, and the overwhelming friendliness of Scots to visitors, to understand why almost half of Scots voters expressed the wish for independence from the rest of the United Kingdom in the 2014 referendum.

4.3 COMMUNITY OWNERSHIP

Participatory Variety Selection (PVS), Village Extension Workers (VEWs) and Village Animal Health Workers (VAHWs) discussed in this author's earlier book exemplify means whereby a community takes ownership of an intervention, so important to assure sustainability of food security. A community forestry intervention is another example, as has proved popular in Nepal for instance, and is regarded by one international nongovernmental organization (INGO) in Myanmar as its most valuable intervention, helping to maintain the forest as the life-giving source of livelihoods that it is, and keeping at bay "land-grabs" from other countries or potentially exploitative concessions to large companies. Without ownership of an initiative, it will last only as long as the project and its financing do. Clearly, there is no sustainability or building resilience in such. Right at the start of any intervention, the concept of community ownership needs to be raised and addressed.

Working *with* people not *for* them is the appropriate attitude for outsiders engaged in an intervention, be they farmers, fisherfolk, teachers and parents in a school meals program, or patients and staff at a CMAM clinic. "Farmer field schools" is a friendly modality in which such equality comes naturally. For "implementing" outsiders to live in the community is a plus point too, rather than arriving in a 4WD each morning straight from a smart urban hotel.

An outsider should not assume that a much-needed water pump for a community should be installed near the village center—the women who would use it may prefer a more hidden place so they can associate and discuss out of sight of their menfolk; the users need to be consulted on its location, rather than the decision being left only to hydrologists or "planners".

A classic example of failure, where a project "worked for people rather than with them", happened in Bundibugyo district, in the western Rift valley of Uganda. A well-meaning charity wished to upgrade the housing conditions of the Batwa forest dwellers, and at the edge of the forest built them a village of new houses with corrugated iron roofs. Some of the Batwa were encouraged to move in and try out these structures. However, when the first tropical rainstorm arrived, the clatter of rain on the metal roofs caused everyone to flee in panic back to the forest, and they were not seen again, despite a ready abatement of the noise problem being at hand—leaf fronds from the forest as roof cladding.

Part of the recipe for ownership to be seamlessly acquired is that to some extent the pace of work has to be acceptable to the beneficiaries. Much of the time, this pace is slower than the outsider would like. At other times, the pace may be faster than the outsider can cope with. Watching people working in their rice fields in The Gambia in June 2017 amazed me. It was early afternoon, the outside temperature I estimated was $35°C$, and it was Ramadan. I had opted to undertake the dawn-to-dusk fast too, as a courtesy to my national colleagues and the community, and the very thought of such physical work under those conditions made me feel weak.

An interesting example of the need to move at the appropriate pace, at which the other party feels comfortable, and to use a carpentry metaphor, to "work with the grain of the wood rather than against it", is offered below. It involves President Arafat's visit to Oslo in 1994 to receive the Nobel Peace Prize, shared with Shimon Perez and Yitzhak Rabin of Israel "for their efforts to create peace in the Middle East". When the organizers deemed it time for Arafat and his party to be taken from their hotel to the ceremony, he and his Palestine Liberation Organization (PLO) party were watching a *Tom and Jerry* cartoon on television. Arafat let the then-Secretary of the Norwegian Nobel Committee, who had come to collect him, know that he and his party would not be going anywhere until the cartoon had finished! Palestinians' "political" engagement with that program will not be lost on the reader, their identifying strongly with Jerry the mouse, who despite his small size always wins against Tom the cat, who is larger and stronger. The Committee Secretary waited…

Certainly during the development process we need also to learn how to let go, building on an "exit strategy", so that the work can continue and expand "post-project". To quote the Bohemian poet Rainer Maria Rilke:

We need, in love, to practice only this:

letting each other go. For holding on

comes easily; we do not need to learn it.

(Mitchell, 1989)

Just like any parent, one needs to give space for the growing child to learn, to learn from his or her mistakes, to learn how to do things better, and equally important, to acknowledge that parents make mistakes too.

4.4 SUCCESS BREEDS SUCCESS

Through each success brought to site, especially in the case of "quick wins", confidence of the community (in itself and the intervention) is increased and any negative attitude can be mitigated. Increased confidence may take the form of accepting greater degrees of experimentation and prepared-ness to take higher-level risks. Ideas that work in practice, as assessed by the participants, are passed on through referral, first to relatives, friends and neighbors, the methodology of continuous improvement being socially in-stitutionalized at the local community level.

4.5 INDIVIDUAL FOOD SECURITY STRATEGIES

The development effort referred to above in this chapter is "communal", yet each individual in the world must first have his or her own strategy of obtaining food, with probably more than one pursued simultaneously by any individual to bolster resilience. Strategies may comprise growing one's own food, earning through employment to buy it from a market, obtain-ing it through a hunter-gatherer lifestyle, bartering another commodity or service for food, being recipient of humanitarian aid, expatriate remittance, government pension or safety net, begging/scavenging/stealing, turning to a family safety net or calling down favors owed from networks.

Regarding this last option, the relationships that people develop, espe-cially in poor communities, in order to secure a stable food supply is an important aspect of the sociology of food consumption. Such networking with others to improve the chance of having food at times of scarcity on a reciprocal basis usually involves relatives, neighbors and/or store owners. The term "social capital" embraces this (McIntosh, 2013). Social capital re-fers to the collective value of all "social networks" (who people know) and the inclinations that arise from these networks to do things for each other ("norms of reciprocity").[15] Social capital relates to the internal social and cultural coherence of society, the norms and values that govern interactions among people and the institutions in which they are embedded. It is the

---

[15]http://www.oecd.org/insights/37966934.pdf (accessed December 15, 2015).

glue that holds societies together, without which there can be no economic growth or human wellbeing. Without social capital, society at large will collapse, and today's world presents some very sad large-scale examples of this. There is growing empirical evidence that social capital contributes significantly to sustainable development. Human and social capital do not exist in isolation from each other; the two are linked in complex ways and, to some extent, feed off each other.

Two examples of social capital in building personal resilience against food insecurity that I have found particularly interesting involve the networking and complex system of social relationships among street children in Addis Ababa, which underlie their survival, work done by my former colleague in Ethiopia for her Master's degree (Asheber, 2005). A second example is the value of ethnicity in providing incentives which enable the flow of resources across generations, thereby eliciting investment and the formation of capital, and growth in developing societies (this positive outcome is the upside of the more commonly discussed ethnic conflict leading to violence) (Bates, 1999).

Some of the colorful individual food security strategies I have personally encountered in recent times are mentioned below.[16]

4.5.1 Sudanese Goatherd

In 2016, while driving across the desert from Port Sudan towards the hamlet of Amour, in the Aulieb zone of Red Sea State, three to four hours' drive away, my party looked for somewhere to have lunch. We were driving up a dry wadi bed and made for the largest tree we could find. No sooner had we drawn up under the tree than an old lady who was lying in its shade stood up and started hollering at us. "She's mad", said one of my Sudanese colleagues. On the contrary, I opined, she was likely the most sane among us. She had a plan. Her day job was a goatherd, and she was there to keep her small stock in order as they grazed on whatever they could find in the wadi. She saw us as a source of supplemental income, claiming that the tree was hers and we had no right to use it. I offered her 10 Sudanese pounds, and she welcomed us. She and her young nephew also ate a good lunch, from the food we had prepared for ourselves but did not eat. In the evening when we passed her at the tree on our return journey, we exchanged hearty waves.

[16]National food security strategies were covered in Section 5.2 of the author's earlier book.

4.5.2 Sudanese Shoe Shiner

I was very moved by a child I came across in Wad Elhilew township near the Atbara dam in Kassala State. My team had stopped its visits to small-holdings for lunch, which consisted of Nile perch freshly caught from the lake behind the dam. While we were drinking tea and waiting for the fish to cook, a boy appeared from the dirt street offering to clean our shoes, which we surrendered to him. I tried to capture his work on camera (Photo 4.1). I never got the shot I really wanted, as his modesty prevented him raising his eyes to meet mine for more than a split second, yet his eyes are forever captured in my memory. His name is Salih Omar, then 11 years old, and he needed the money not just to buy his own food, but to feed his siblings and parents, and to buy exercise books and pencils for his *madrasa* (a school/college for Islamic teaching).

Photo 4.1 Salih Omar, shoeshiner, Wad Elhilew township, Kassala State, Sudan (April 27, 2016).

4.5.3 Fishing Community

Life must be tough for women in a fishing community on the Red Sea coast of Sudan. I divine this because while I was visiting a women's group in one such community on an assignment in 2016, no sooner had I entered the group's meeting room than I received three proposals of marriage—according to my Sudanese colleagues. The women were half my age, but this was their Plan B food security strategy. Whenever I enthused to my colleagues about this fishing community and the project intervention we had come to evaluate, they did not spare me their taunts at a possible alternative reason for my wanting to see more of it!

4.5.4 Tomato Growing in the Sudanese Desert

Well-irrigated Beto 86 tomatoes perform well despite the heat in the desert of Red Sea State, and the fruit has a sufficiently robust structure to enable it to travel to market with minimal loss of quality. However, marketing from the farm in Kurdeib (Photo 4.2) is still the main constraint, as access to Port Sudan is via tracks across rock and coarse sand hilly terrain. A market intermediary provides a lorry, which takes some three hours, and the "profit margin" is slim. Grass "weeds" which flourish among the tomatoes

Photo 4.2 Hashim Omar Eisa, local clan leader, surveying his three *feddan* (1 feddan = approximately 1 acre (0.42 hectares)) of tomato fields at Kurdeib, Gunub Zone, Red Sea State, Sudan, which had sustained hailstorm damage 3 days previously (April 22, 2016).

are harvested as a valuable fodder for goats in the evenings (see Box 8.2 in Chapter 8). This initiative is overseen by the NGO SOS Sahel, under the Sudan Food Security Program, for local households interested in horticulture to supplement their diet and main livelihood of goats and rainfed sorghum (using rainwater harvesting techniques).

4.5.5 Sale of Fodder in Sudan

In much of northeastern Sudan, Sahelian conditions pertain. This leads to livestock fodder being an item of trade, especially sought after at the end of the dry season and when rains fail. Photo 4.3 shows three camels just short of reaching their destination of Hamish Koreib, in Kassala State. The lead rider explained that the round trip from home to market takes 7 days. On this outward leg his camels are carrying 50 "bunches" of dry Doum palm leaves for coarse camel fodder, each of which will sell in Hamish Koreib bazaar for two Sudanese pounds (SDG), thereby grossing 100 SDG (equivalent then to about US$7.5, black market rate). With this money the travelers will buy grain, salt, coffee and sugar before the trip home (see front cover photograph).

Photo 4.3 Doum palm fronds as an item of trade, Hamish Koreib, Kassala State, Sudan (April 28, 2016).

4.5.6 Police in Sudan

At dusk on the road from Khartoum to Port Sudan, my vehicle was stopped at a police roadblock, comprising a red and white counter-balanced barrier. Three people were involved in a well-oiled small-scale enterprise, which worked as follows. A junior police officer approached our vehicle, and informed our driver that it was not possible for us to proceed. The driver remonstrated yet the police officer would not budge. Next, a senior officer dismounted the barrier and approached the vehicle. He assured the driver that all would be well if we could purchase some sesame biscuits from the third person in the enterprise, who had been sitting on the barrier next to the senior officer, and was perhaps his girlfriend. She now approached the vehicle with her basket, and there was an exchange of cash and biscuits. The junior officer then reappeared from the shadows and lifted the barrier, whereupon we proceeded.[17]

4.5.7 Police on the Owen Falls Dam, Uganda

As my pick-up approached the dam on the Nile at Jinja I noted a police officer set off across the bridge on his motorcycle. He drove barely quickly enough to avoid falling off his bike, so I overtook him. As soon as I had exited the bridge, the aforesaid officer overtook me and pulled across so I had to stop. He dismounted his bike and reached for his notebook. He said I had overtaken him on the bridge, and if I were to do that to a car I could cause an accident. I assured the officer that I had never overtaken a car on that bridge, not in the 40 years during which I had been driving across it. Presumably fearing that due to my obvious familiarity with the country I might know people who could cause him trouble, and put an end to his moonlighting income stream if not his primary career too, he put away his notebook and waved me on. His supplementary food security strategy had failed on that occasion.

4.5.8 Police on the Fort Portal Road, Uganda

In 2012, I was in my pick-up carrying 5,000 tree seedlings in plastic sleeves, to plant on my family farm. A policeman stepped out from the shade of some trees between Kampala and Mubende and waved me down. He lifted the papyrus mat with which I had covered the seedlings and inspected

---

[17]I considered including this anecdote under Section 7.2 later in this book, "Getting it right as a team", as the indirect objective of the enterprise was sound—to increase the demand for nutritious sesame and hence its local production!

the cargo. He then pronounced that the seedlings had been loaded in a dangerous way and I had therefore broken the law. "No officer", I replied, "the seedlings were loaded by staff of the Department of Forests, who have been doing this work for ever". While the officer thought up another line of attack, my driver got out of the cab and made his way to the back of the vehicle, the police officer in close pursuit. Within seconds, the driver was back in the cab and we proceeded. A small financial transaction had taken place out of my sight.

4.5.9 Room Cleaner in The Gambia

During an evaluation mission in The Gambia, I made my Banjul base a cheap and cheerful backpacker's guest house right on the sea shore at Farafara. I came to know that what I earned in a day as a consultant was more than my room cleaner could earn in a whole year, 6 days a week. Her income just about covered the rent of the accommodation she shared with her husband in a poor part of town … for part of the year anyway. During the tourist season, when Gambians from upcountry descend on Banjul hoping to share in the spoils, the room rent rises and she has to move to an even cheaper room on the edge of town, together with her furniture. When her jobbing builder husband manages to find work, the couple can afford to eat as well as pay for their shelter. She could not understand when I said that I was interested to know more about her life. That *was* her life—what was "interesting" about it?! The couple had lost both of their children soon after birth, seemingly related to the mother's poor nutritional status. I set up a bank account for her with one month's rental as a contingency, yet what a pitiful response that was in the face of her life's challenges.

4.5.10 Private Enterprise in Refugee Camps

Even under very harsh conditions, private enterprise may prosper, as in the case of a bakery business set up by women in a Palestinian refugee camp in Jericho, West Bank (Photo 4.4). Similarly, in 2011, there was an exodus of refugees from the Ivory Coast, after the losing candidate in the presidential election, Laurent Gbagbo, refused to concede defeat. Many of these refugees crossed into northern Liberia where I was working at the time, and I noted that as soon as the UNHCR tented camp started to fill, petty traders were on the ground selling essential items to the incoming Ivorians.

Photo 4.4 Ramadan biscuits being prepared at the Women's Development Centre Cooperative, Jericho, Palestine. Director Samar Moyasar is on the far left (June 26, 2016).

REFERENCES

Asheber, T., 2005. Social Capital as a Survival Mechanism: The Case of Some Street Children and Youth in Addis Ababa. (M.A. thesis) Department of Regional and Local Development Studies, University of Addis Ababa. June. 138 pp. http://etd.aau.edu.et/bitstream/123456789/1790/3/Tsedey%20Asheber.pdf (accessed September 12, 2015).

Bartlett, P.F. (Ed.), 1980. Agricultural Decision Making: Anthropological Contributions to Rural Development. Academic Press, New York. 378 pp.

Bates, R., 1999. In: Ethnicity, capital formation and conflict. Social Capital Initiative Working Paper No. 12. September. World Bank, Washington, DC. 47 pp. http://www.worldbank.org/socialdevelopment (accessed September 12, 2015).

Berry, S., 1980. Decision making and policy making in rural development. In: Bartlett, P.F. (Ed.), Agricultural Decision Making: Anthropological Contributions to Rural Development. Academic Press, New York, pp. 321–335. (Chapter 13).

Cancian, F., 1980. Risk and uncertainty in agricultural decision making. In: Bartlett, P.F. (Ed.), Agricultural Decision Making: Anthropological Contributions to Rural Development. Academic Press, New York, pp. 161–176. (Chapter 7).

Chibnik, M., 1980. The statistical behavior approach: the choice between wage labor and cash cropping in rural Belize. In: Bartlett, P.F. (Ed.), Agricultural Decision Making: Anthropological Contributions to Rural Development. Academic Press, New York, pp. 87–114. (Chapter 4).

Cohen, P.S., 1967. Economic analysis and economic man: some comments on a controversy. In: Firth, R. (Ed.), Themes in Economic Anthropology. Tavistock, London, pp. 91–118. (p. 104, Association of Social Anthropologists of the Commonwealth).

Devische, R., 1993. Weaving the Threads of Life: The Khita Gyn-Eco-Logical Healing Cult Among the Yaka. University of Chicago Press, Chicago. 334 pp.

Drèze, J., 2017. Sense and Solidarity: Jholawala Economics for Everyone. Permanent Black, Delhi. 353 pp.

Gladwin, H., Murtaugh, M., 1980. The attentive-preattentive distinction in agricultural decision making. In: Bartlett, P.F. (Ed.), Agricultural Decision Making: Anthropological Contributions to Rural Development. Academic Press, New York, pp. 115–136. (Chapter 5).

Johnson, A., 1980. The limits of formalism in agricultural decision research. In: Bartlett, P.F. (Ed.), Agricultural Decision Making: Anthropological Contributions to Rural Development. Academic Press, New York, pp. 19–43. (Chapter 2). p. 41.

Kent, S., 1996a. Cultural Diversity Among Twentieth-Century Foragers: An African Perspective. Cambridge University Press, UK. 344 pp. p. 16.

Kent, S., 1996b. Hunting variability at a recently sedentary Kalahari village. In: Kent, S. (Ed.), Cultural Diversity among Twentieth-Century Foragers: An African Perspective. Cambridge University Press, pp. 125–156. p. 154 (Chapter 6).

McIntosh, W.A., 2013. The sociology of food. In: Albala, K. (Ed.), Routledge International Handbook of Food Studies (2012). Taylor & Francis Publishers, Abingdon, pp. 14–26. (Chapter 2).

Mitchell, S., 1989. (trans.) The Collected Poems of Rainer Maria Rilke. Penguin Random House, London.

Rose, D., 1990. Living the Ethnographic Life. Qualitative Research Methods Series 23. Sage, Newbury Park, CA. 64 pp. p. 12.

Scott, S., 2015. Negotiating Identity; Symbolic Interactionist Approaches to Social Identity. Polity Press, Cambridge. 263 pp.

CHAPTER 5

The Starting Point of a Development Intervention

Contents

5.1 Introduction 49
    5.1.1 Food-Sourcing Context 49
    5.1.2 Approach to Improving Resilience 52
    5.1.3 Expatriates—Blessing or Liability? 56
5.2 Personal Journeys to Our Understanding of "Food Insecurity" 58
5.3 Seeking Consensus 61
5.4 Challenging One's Assumptions 63
    5.4.1 The Need to Triangulate 63
5.5 Interaction With Local Administrations 66
5.6 Lack of Trust Within Multiethnic National Communities 66
5.7 Conflict- or Political-Break With Tradition 67
5.8 Managing Expectations 68
References 69

5.1 INTRODUCTION

5.1.1 Food-Sourcing Context

The major food-sourcing strategies in the world have been neatly summarized by Nanda and Warms (1998) in a book which brings cultural adaptation into focus. These strategies comprise hunting and gathering, pastoralism, extensive agriculture and intensive agriculture, across six major environmental zones, each with a particular climate, soil composition and plant and animal life.

1. *Grasslands* (may be called savanna, steppe, prairie), which can provide livelihoods for hunter-gatherers and pastoralists. Grasslands cover some 25% of the earth's land surface and support 10% of the world's population.

2. *Deserts* of various types, covering some 18% of the earth's surface, and supporting 6% of the world's population. These can sustain hunter-gatherers, and intensive agriculture and dense human populations if water sources are exploited at oases and/or using deeper aquifers.

3. *Arctic and subarctic zones* cover 16% of the land surface, supporting less than 1% of the world's population, who live by hunting, trapping and herding to an ever-reducing degree, with more living in modern settlements, earning from mining and other activities.

4. *Mountains* cover 12% of the land surface, supporting 7% of the earth's population, with livelihoods relating to extensive agriculture and/ or pastoralism.

5 and 6. More than three-quarters of the world's population are supported on the remaining land surface, *nominally tropical or temperate forest* (Western Europe and Eastern Asia especially), though many of the forests have been cut, and the land used for extensive or intensive/industrial farming activities.

In much of the developing world, the rainy/temperature seasons determine the availability of resources for food producing/gathering activities, which in turn requires adaptation of producing and gathering strategies. The strategies also have to be adapted in response to unpredictable short-term changes in the environment (droughts, floods, disease and pest outbreaks for humans, crops and livestock). The smartest adaptations to environmental change and uncertainty are those which are wide-ranging, attuned to a variety of resources. Hunter-gatherers have a great diversity of food sources, so when some are not available, others are.

A given society can also extend its resource base through trading with others. Exchange of goods and services is common between pastoral and settled agricultural groups in northern Nigeria, for example, though this needs to be well-managed. Wilkie (1988) gives an example from the Ituri rain forest in Central Africa in which several groups of Mbuti foragers have complex hereditary exchange relationships with the Lese people, their agricultural livelihood neighbors. The Mbuti bring meat, mushrooms, honey, building materials and local medicines to the Lese, and provide labor on the Lese farms. In exchange, the Mbuti receive cassava, green bananas, groundnuts and rice, which contribute more than half of the Mbuti diet. The Lese also provide the Mbuti with metal to fashion their hunting instruments, aluminum cooking pots and cotton cloth. The Ituri forest of DR Congo overlaps the border with Bundibugyo district in the western Rift Valley of Uganda, and I have myself been offered forest products by Batwa forest dwellers, when I have been in Bundibugyo township.

The major global food-gathering strategies are hunting and gathering (foraging), pastoralism (transhumant and nomadic), extensive agriculture (using non-mechanical means), intensive cultivation using animal draft plow, and the more heavily-mechanized industrial-level, high-input-high-output agriculture. Each type of food-sourcing has some social correlates, in terms of organization and cultural values (though there is great variability).

For most of human history, foraging has been the major food-getting strategy. As tools improved, agriculture became possible, though in the arctic, climate and terrain have prevented agriculture from developing. Foraging strategies may be more dependable than cultivation in places where both could be followed. Agriculture developed to bring greater productivity per unit of land and higher efficiency (yield per person per hour of labor invested).

Some hunting populations made the transition to cultivation very recently, much preferring the economic, social and psychological benefits of a foraging or pastoral life. In these societies, hunting and pastoralism are highly valued and intimately connected with a people's cultural identity. Clashes between pastoral and agricultural communities can occur, when a land or water resource is shared and there is insufficient to go around. In a typical non-industrial society, over 80% of the population is directly involved in food production, whereas in a highly-industrialized society, 10% of the people directly produce food, both for themselves and for the other 90%. The environmental and social destruction associated with modern global economics has led to a reactivated interest and respect for the ways nonindustrialized peoples have adapted to their environment, and realization that there could be value in trying to incorporate some of the old ways (see Sections 6.1, 11.1 and 8.9.1).

Employing indigenous technologies can lead to the sustainable holistic exploitation of forest resources, for example, for food, medicines and building materials. Logging without replanting can transform tropical forest into wastelands, through plunder for short-term economic gain, as has been done in Amazonia, Liberia and The Gambia, for instance. While modern technology and scientific knowledge can transform deserts into gardens, gardens can revert to deserts if worked without due regard to sustainable exploitation of natural resources. Introduction of domestic livestock destroyed the carefully-managed agricultural ecosystem of the Inca Empire in South America, and the European demand for sugar and tobacco resulted in huge areas of monocrop agriculture that transformed the land cover of southern United States, which in turn gave rise to the introduction of African slavery (Scammel, 1989; Mintz, 1985).

5.1.2 Approach to Improving Resilience

Economic development programs addressing "community development" are based on the rationale that people can improve the material and non-material conditions of their lives through community autonomy and self-sufficiency. One underlying assumption in the past has been that the behavior, values and beliefs (viz. cultures) of local communities engaging in often-subsistence economies are resistant to change and that such resistance is the primary obstruction to economic development. Government and development agencies have aimed to change behaviors of local communities so they become more economically productive, and through changed attitudes develop the competence and confidence needed to better help themselves. Ideally, a community can realize its own goals by using its own human resources while recognizing that some outside support, both financial and technical, will be needed along the way. There is a need to discover the felt needs of the population at local level and work through these to achieve improvements. This is more complex than it may seem, as local communities are not homogeneous and development will not benefit the whole community. Interventions need to be culturally-sensitive, otherwise development projects will fail. The "local community" is not a closed culturally-stable system, however, as it is likely increasingly influenced by outside forces and structures. Also there are often conflicts within communities (Nanda and Warms, 1998, p. 359). Box 5.1 sets out the framework for an "outsider" to understand resilience.

Anthropologists have helped development agencies understand that economic, social and other decisions made by local communities are not merely a matter of cultural traditions but rather are usually rational and necessary, given available economic, human and environmental constraints. For instance, the efficiencies of patterns of pastoral livestock management over a long period must be recognized.

Any attempt to help people from the bottom-up must also take into account the larger political and economic structures and processes that limit people's access to beneficial opportunities. Development anthropologists endeavor to recognize the importance of power and coercion which affect communities, and try to understand and manage them. Previously, the failure of communities to solve their own problems was assumed to be due to lack of motivation and organization. In fact, it may be ascribed to lack of natural resources and national bureaucratic mismanagement and corruption. Those who used to exercise local power were local people elected based on their wisdom and the trust that the community had in them. That was how

BOX 5.1 Understanding Resilience

Resilience may be framed in terms of three capacities of the target group, country or region—absorptive, adaptive and transformative, as formulated by the Organisation for Economic Co-operation and Development (OECD). Improving such capacities needs to be the guide for designing and implementing programs which seek to combat and remove food and nutrition insecurity in perpetuity:

1. *Absorptive capacity*—the capacity to cope with future shocks and stresses by preparing for, mitigating or preventing negative impacts, using predetermined coping responses in order to preserve and restore essential basic structures and functions. Coping mechanisms include, for example, early harvesting, taking children out of school and delaying debt repayments. That such future shocks and stresses will occur is uncontestable. Only their nature, timing and sequence are variables. Causes may be related to natural disaster or man-made disaster (such as climate change or conflict).

2. *Adaptive capacity*—the capacity to make adjustments and incremental changes in anticipation of, or response to, change, in ways that create more flexibility and optimal outcomes in future. This includes capacity to take advantage of opportunities, for example, through diversification of livelihoods, involvement of the private sector in delivering basic services, and introducing seeds of drought-tolerant "varieties".

3. *Transformative capacity*—the capacity to challenge and reverse overarching "systems" (such as poverty) that currently increase risk, vulnerability and inequality, by addressing drivers of these components. However, systemic change takes time and sustained engagement.

it was when I first went to live in Uganda, shortly after Independence in 1962. Now, local government comprises appointed officials, some of whom have no roots in the local community. When there are locally-elected officials, they are likely to have achieved their posts through bribing voters in return for giving promises they cannot fulfil. They then charge for their services to recover their initial outlay. The two sets of officials work hand in glove, and at times it seems almost impossible for local people to wrest control of their lives from them.

Many things need addressing before a development project intervention is made, including conducting a stakeholder analysis and engaging with representative community leaders. Even prior to interacting with people on the ground—a "windshield survey", as my American friend calls it—can be instructive. In 2011, on my first visit to Liberia, I learned a lot driving from Monrovia to Buchanan and back. I noted, for example, the virtual absence of goats, something unthinkable in any other country in Africa in which I'd lived for 25 years. This surprise had as its explanation the quite recent long-running civil war in the country.

Community development requires teamwork, yet each one of the team has a different *starting point* when setting out on a project to improve resilience, depending on upbringing, education and experience in the world. That determines what we see and hear (and the other three senses too— smell, taste, touch), and, just as important, what we "understand" of what we see and hear, and not least, what we "feel".

The starting point shapes and conditions who and what we are, and how we integrate with others. During my work across some 30 developing countries, I have had ample exposure to development-related instances underlining how experience makes each of us different from all others (a composite layer superimposed on genetic/physiology-based differences). The combination explains our various levels of single-mindedness, motivation, commitment and ability to engage for instance (Laming, 2004), and our awareness of the conditioning factors on resilience against food and nutrition insecurity. The female cab company owner in Zambia I met whose husband had been eaten by a crocodile; my colleague in Uganda from Zimbabwe whose wife had been killed at a roadblock during the independence struggle; a girl I met in Uganda who had escaped from captivity by the Lord's Resistance Army (LRA) in northern Uganda; my office secretary in Maiduguri Nigeria whose husband had been implicated in a failed military coup, and executed—each of these is a witness to the world, its threats and dangers, through a different lens, because of their varying experiences being threaded into the fabric of group social and cultural norms.

The meaning of "bus" or "train" is rather different for someone living in London compared with someone in India, say—in terms of both its appearance and riding on one. Similarly, the concept of "street food" conjures up disparate expectations and experiences in the two locations (Photos 5.1 and 5.2).

Reading about mountains is a far cry from living in them. The latter is a transformative experience, learning of the conditions over the year, the lifestyles of people, and why they do some things and don't do others. Having lived for five years on the lower slopes of the Himals in Nepal, showering in near-freezing water under the hand pump in the garden, I soon realized why children in remote communities there have lice in their hair through not having it washed more frequently.

Photo 5.1 Street food seller in Kolkata, India (April, 2013). *(Courtesy of Martin Tayler.)*

Photo 5.2 City street food in Amritsar, Punjab, NW India (August 2014). *(Courtesy of Martin Tayler.)*

Another experience, in the mountains of Ethiopia, was one of the most miserable nights of my life, in which I learned a lot about the physiology of my own body. Wishing to get from Bahir Dar, the capital of Amhara Region, to a valley in Wollo province, my driver decided on a short route along a one-lane mountain track hacked from the rock face. As luck would have it, soon after starting our descent on the far side of the massif we were enveloped by low cloud, and the driver refused to move, saying we would stay overnight there and continue when the cloud lifted in the morning.

The mist turned to rain, with the wind so strong it shaped each raindrop into a dumbbell. I had not brought warm clothing, as we had started the journey in baking conditions. Throughout the long evening and night in the vehicle I alternately shivered because of the cold, then had to breathe deeply after each bout of shivering to replace the oxygen used in the blood. At daybreak the rain stopped, the cloud dispersed, and we proceeded down the mountain track, arriving half a day late at our destination.

In Africa I learned the generosity of the Kamba people in Ukambani in dryland Kenya, who would donate one of their few chickens to the driver of our vehicle to say thanks that we had visited their homestead; the generosity of people in Borno State in Nigeria when they knew my group had just been robbed at gunpoint by Boko Haram, providing free accommodation for us in the hotel; and the countless occasions when I have been welcomed into remote tented communities in the Sahara, and feted as if I were a long-lost relative. These experiences are indelible in my memory, and I give thanks to all those people whom I'll never meet again who offered their friendship to us strangers, and sacrificed for us out of their generous spirit.

On the downside, experience is a great teacher from which one can learn about the worse side of human nature—such as negotiating military roadblocks in a country ruled by a brutal dictator, and corruption in national and local government leading to contract awards being determined not by worth of the bid but worth of the associated "brown envelope". With each new experience comes a new "starting point", at a higher plane of understanding of the range of human behavior, and its importance in determining development outcomes.

5.1.3 Expatriates—Blessing or Liability?

These different starting points can be particularly challenging when expatriates are concerned, sometimes based in donor office agencies in major cities, making decisions on which intervention to fund in places far away that they have little idea about. This is exemplified by an experience I had in Lusaka, Zambia, when I heard a member of a visiting expert team which had come to conduct an evaluation say, "We'll hold discussions in Lusaka, then go to the field if we have time". In my view, the priority should have been the reverse if a choice had to be made. Yet both are needed, a visiting team not relying alone on secondary information from donor and government offices, itself perhaps gleaned by remote control.

I have come across a wide range of competencies among expatriates, some bringing their values and way of doing things from their home country, and seemingly unable to adapt and relate to local conditions, competencies and aspirations. Youngsters in particular are prone to getting it wrong many a time at the start of their careers, and I count myself among these. "Did you make those terraces for them?" a young visitor asked me, confronted with the earthen rice terraces cascading down the lower hills in eastern Nepal, unaware that they took many generations to shape as a rational form of land use. Another young visitor to the towering Separation Wall in Bethlehem said to me, "I can't understand why the Palestinians built this wall". I replied, "If you do some research on it, I think you'll find it is not like that".

Yet the worst-case scenario is when the expatriate remains ignorant despite his or her exposure to developing countries, and occupies a position of authority in a country donor office. I've sometimes seen nationals in the donor office held hostage in such situations, wishing to speak out, yet mindful if they do that they will lose their positions and the income on which they and their extended families depend.

By contrast, other expatriates (hopefully the great majority) are knowledgeable about their new work station, bringing massive and varied international experience and wisdom to share with their counterparts and other development partners. Nationals may even find themselves surprised at how well an expatriate can paddle a dug-out canoe, for instance (an adapted skill from childhood exploits on seaside boating lakes) and the stamina with which he or she can trek for days in mountainous country with a heavy backpack (skills learned at school in scouts/guides or in a cadet force), and so on.

I would counsel that each expatriate going to a country first take the trouble to learn about the country in which he or she intends to work. Most importantly, he or she should understand the people therein, not just the geography and climate. For instance, for an outsider to understand the people of Myanmar, I would suggest that as well as reading a travel guide produced for tourists telling them about the weather and listing the pagodas to visit, it would be appropriate to buy a totally different sort of book. The one which my (economist) elder son Austin thoughtfully bought me before I visited that country was "Golden Earth: Travels in Burma", published back in 1952 by Norman Lewis (described by novelist Graham Greene as "one of the best writers, not of any particular decade, but of our century"). From this book I came to understand so much about how the people thought and behaved then and *now*, about interethnic tension, for example, which I was able to test during my own interactions with the people. Lewis manages to get right into the psyche of the Burmese, because he listened and observed, and continually asked himself, "Why?". Having reached Myanmar, I then set about ground-truthing my understanding of Lewis' tenets, finding that they all held up to scrutiny 65 years later.

Once the expatriate truly empathizes with the national, then the stage is set for him or her to contribute fully, by helping to bring the best of what is new to the best of what is old (see Section 8.9). To start with though, he or she will flounder, as despite likely wide experience elsewhere, he or she won't know where to get "widgets" made locally! An international also has utility value as he or she is able to avoid pressure from local bigwigs which local consultants may find more difficult, as they will likely need to maintain cordial relations with that bigwig for future employment prospects, *inter alia*. The expatriate can also disagree with a Minister in private or public debate, which a national could not contemplate doing if he or she expected to retain employment, even to stay out of jail. Military States or States run by fear I particularly have in mind.

5.2 PERSONAL JOURNEYS TO OUR UNDERSTANDING OF "FOOD INSECURITY"

Each of us has been on a personal journey leading to where each of us is "now" regarding our specific understanding of food security and resilience against it. My own journey began in Bristol after the end of the Second World War. That English city had been heavily bombed, especially during the Bristol Blitz from November 24, 1940 till April 11, 1941. During my childhood, my parents rarely spoke of the war in my company. I noticed that every 20th house on my street was in ruins, and overgrown with the primary colonizer bush *Buddleia*, which was a source of wonder for me as I walked to school, as during its summer flowering it attracted beautiful butterflies. I didn't question this order of things, or why the houses had "collapsed". Only later did I come to know more about the war and that my parents' house had been one of those bombed and they were lucky to have survived.

I recall the beans on toast as a recurrent meal, with tripe and onions as a Sunday treat. I recall too the food ration coupons, and the child allocation of orange juice and cod liver oil from the government. At the top of the road was a free-range chicken farm. By the age of five I was tall enough to see over the boundary wall on my way to play in the local park. I noticed that one deviant chicken laid her eggs beneath one of the wood slat chicken houses, rather than inside the house as did the other chickens. When no one was about, I would climb over the wall, pocket the eggs, climb back over the wall and take them to my mother. Though I noted how pleased my mother was when I brought her the eggs, it was not until decades later that she told me that as eggs were unavailable or unaffordable for her in the shops after the war, without the eggs I brought home she would have had none to cook for me.

More than 50 years later I made friends with the Minister of Agriculture in Palestine, and he told me that post-1948 when his family had fled historic Palestine to become a refugee in Jordan, he too engaged in similar nefarious activity when a child. He would play with children of police officers at the police station, and noticed a wayward chicken which laid its eggs in the bushes. When no one was looking, he pocketed those eggs and took them to his mother. That shared experience of a strategy to combat hunger and poverty at home consolidated our friendship. We later collaborated on formulating the country's Food Security Strategy, endorsement of which he steered through Cabinet.

My second experience of what warfare means was on Salisbury Plain in England, a training ground for the British Army. I was 14, a school cadet in the Gloucestershire regiment, and we were on a "night exercise". All was quiet under the stars, until the Lieutenant gave an order to his Bren gunner lying on the grass, who let rip into the top of a nearby tree. The sudden deafening clatter was one of the most shocking things I've ever encountered. Soon enough though there followed 40 years of my working in conflict-prone and post-conflict countries in which there have been many experiences which pinpointed the personal insecurity challenge to nationals trying to remain food-secure, involving military coups, near-misses from suicide bombs, imprisonment by a dictator, witnessing the lives of countless street children, and having my vehicle burnt by "Maoists" in the hills of Nepal. To these experiences can be added countless stories heard from others, showing the food security fragility of each of us.

Living in **Yemen** under the Ali Abdullah Saleh regime showed me that that country (especially the north, the former Yemen Arab Republic) was an accident waiting to happen. Citrus orchards are being grubbed out and replaced by the narcotic plant *Catha edulis* ("khat"), and irrigated by fossil water from the falling aquifers, most adults high on chewing that drug in the afternoon and evening rather than thinking of how to learn new skills and improve future prospects for their families and country. Similarly, work-ing in **Libya** under Colonel Gadhafi was discouraging, as he had told his people that they didn't need to work, for now it was Libyans who were in charge, not the old colonial power ... it was difficult for foreign experts to transfer their skills to nationals, as the incentive for that to happen had been undermined by the political directive.

In **Afghanistan** I was not surprised to learn that the Afghan doctor treating me for a virus infection in the German field hospital in Kabul did not have enough money to buy medicine to treat his mother's Leishmaniosis, as his small salary was all allocated for basic food, rent and utilities; that one of my field assistants in my Ministry was a medical doctor, yet who earned more on my project than he did in the government hospital, so had left his work there; not surprised either that my national consultant in **Sudan** earned as much during his 25-day consultancy on my donor-funded proj-ect as he would earn over five and a half years in his government job, from which he had taken a short leave; that in Kebbi State in NW **Nigeria**, where cattle keeping is common, children of poor landless people living on the edge of its major town Birnin Kebbi are stunted, because although milk is available in town, their families cannot afford to buy it. My journey over the years has had me living in a traditional wooden house in the west-ern hills of **Nepal**, woken each morning at 4 a.m. by the sound of village women grinding their grain.

I'm familiar with the slums of Kampala, **Uganda**, and the frequency of kwashiorkor among slum children fed on a diet of little else but protein-poor *matooke* (cooked green) banana and cassava, the cheapest staple foods avail-able in the market. Those slums were not something I had learned about in a library or lecture theatre, or by watching a *National Geographic* DVD. I've been in and out of slums for decades, going there to get my car fixed, buying my fruit and finger millet flour in their markets, sometimes sleeping in their mud and wattle "hotels" during field assignments. I've grown used to their presence, their reality. Yet when I wrote the case study on urban slums in the developing world for my earlier book, I learned something about myself, my "starting point".

These slums were part of "our" life in the tropics. I knew they were there, knew the reasons why they were there, knew the conditions of the people. Yet I hadn't emotionally engaged with the fact that they should *not* be there, and that something must be done to replace them. And "replace" not in the way that a former mayor of Nairobi city approached the prob-lem, by sending in bulldozers without first finding alternative housing for those whose homes were to be demolished; rather, by addressing the socio-economic conditions of the population which led to their existence in the first place (the "circumstance" Sridhar refers to in Section 6.2.2). I admit to having had a compartmentalized tunnel-vision in my early years in Africa, as I was addressing another area of living conditions: freedom from hunger, by improving agricultural productivity of food legume crops, so that more of their grain could be availed cheaply to the market place to relieve the undernutrition I saw in both urban and rural environments. Bad housing, its sanitary and water supply challenges, and poverty were for my development colleagues in those disciplines to handle.

5.3 SEEKING CONSENSUS

Any intervention needs to start from the place at which the given community is, not the place at which an outsider involved in making the intervention may be. If there is going to be change for the better, there must be a shared and agreed understanding of "where we are now" and then "what is possible, where, with whom and when". Moreover, the "what" and "how" of that intervention must be appropriate to the values and cultural norms of that community, as well as being technically and politically possible, socially acceptable and beneficial, and cost-effective. This debate must take place at the personal and community level, from project-identification onwards. Yet players are often unequal, living and thinking in different worlds far apart, as was discussed in Sections 5.1 and 5.2.

There is many a slip along the way for those outsiders in the development field who do not trouble to learn the ways and thinking of the partner communities with whom they are to work. It can be fun learning, especially if one is humble—the sort of humility shown by Abraham Lincoln when he went to the home of one of his generals in the American Civil War to congratulate him, only for the servant of the General to bring a message that the General was too tired to see the President. The entourage of the President was shocked, yet the President had the grace to tell them, "I will hold the General's horses for him, so long as he continues to bring us victories".

Very often there are excellent development initiatives already in place, and a worthy goal of a project, if the community agrees, is to help upscale these. An example comes to my mind in Dang District, part of the lowland *terai* (plains) of Nepal, involving biogas and livestock fodder. Household biogas has become popular there, and the sowing of the fodder plant

Stylosanthes on the floor of *sal* (teak) forests, each initiative starting with an individual or project champion at village level (see Chapter 10).

Realistic assessments of success, assumptions and risks need to be discussed upfront with the community, together with the latter's obligations as well as benefits to be accrued. Concepts of maximizing impact and outcome, cost-effectiveness and sustainability need to be shared with community leaders, and indeed start with them. This is a time-dependent process that must be seen to be transparent and not rushed. Perspectives of the community need to be gauged in an inclusive gender-sensitive way, so that ownership of the intervention is achieved from the outset (Section 4.3). Anthropological art and craft needs to be attuned to the highest degree here, so that widely-supported opinions of the "beneficiaries" are seen to be endorsed by project team members, and incorporated into the design and implementation phases. The opinions of various levels of the community need to be accessed, including traditional and religious leaders, as well as public sector and legislature representatives. Local opinion leaders will be identified during the debate.

In interactions between project staff and the target community, the facts can be established using various tools, such as SWOT analysis, transect walks, focus group interviews and one-to-one interviews. The questions need to be framed correctly. I give a tongue-in-cheek example from my own experience in the hills of eastern Nepal in the late 1980s—my heavy-laden trekking team was returning to base in Dhankuta from one of our field visits in the hills and stopped at a rustic tea shop for refreshment. We drank all the tea that was in the kettle there. Shortly after we departed and continued our journey, we were overtaken by a youthful pair from the same project, who complained that the tea shop in which they had intended drinking tea (the one we had just left) said it had no tea. "What question did you ask?" I enquired. "Do you have tea?" was the reply. I explained that we had drunk the last of their tea. Had they asked, "Will you make us tea?", they too would have been able to refresh themselves.

Arriving at a commonly-agreed starting point to build resilience against food and nutrition insecurity can be a challenging exercise, particularly for expatriates in the change-management team. Wise council will likely be needed by nationals on the team to ensure that promising new ideas that are brought to the field sit easily with, or replace, the old ones that have worked hitherto (for better or worse) in the community concerned, and are appropriate to the values and cultural norms of that community. Reliable qualitative and quantitative data on livelihoods need first to be sought, ground-truthed and triangulated, understood and internalized. As part of this process, there is need for all of us involved (from both developing and developed countries) to challenge our assumptions.

5.4 CHALLENGING ONE'S ASSUMPTIONS

5.4.1 The Need to Triangulate

The way to determine "the truth" often involves triangulation, through seeking an item of information in a variety of ways, not just through interview. There is no particular reason why anyone should tell another person the truth after all, certainly not the whole truth, especially if that other person is an outsider. Just consider Facebook entries, with a spun account of an individual's life and doings.

Some of us are more reluctant than others to challenge our beliefs and norms which we hold dear. Some notional examples, not necessarily directly related to development issues, are listed below to exemplify the wisdom of challenging our assumptions or what we are told.

5.4.1.1 Sudan

In the Sudanese desert near Port Sudan in 2016, my group of consultants interacted with some farmers growing vegetables under well irrigation. However, I did not understand much of the main discussion between my national colleague and the farmer's leader. After the interview my colleague told me that the "truth" was probably the opposite of most or all of the answers he had been given by the leader!

A further example from Sudan I learned long ago, when I was writing up my Ph.D. in London. Another postgraduate student in the same hall of residence was a Sudanese doctor. I asked him to tell me the biggest difference between general practice medicine in London and rural Sudan. He pointed to the differing levels of explanation of the ailment forthcoming from the patient. "In Sudan", he said, "the patient will say he has a pain in the head", yet that represents a catch-all for hundreds of conditions, and is little or no help to the doctor using normal questioning. Only by enquiring in roundabout ways may the doctor close in on the nature of the ailment.

5.4.1.2 Somalia

While I was flying low up the Somali coast from Nairobi to Hargeisa in a small ECHO plane in 2011, through the window appeared what seemed to be a wide sandy beach, something I had not previously known about. It had not been there when I was on the ground in that part of coastal Somalia in 1986. How attractive for tourists I thought, but for the current security situation down there. On closer inspection, however, the beach was in fact a banana plantation inundated by sand during the "Indonesian" tsunami of 2004, which in Somalia alone caused some 300 deaths, and some 5,000 displaced.

5.4.1.3 Zambia

The 3-year-old daughter of my American colleague was not communicating with her mother, who was becoming increasingly worried that something was wrong with her. One day the little girl could not be located in the evening and her mother went in search of her. She found one of her daughter's friends who had just seen her, who reported, "She said she was going to see So-and-So…". "But she can't speak!" my colleague responded. It became clear that the toddler could speak the local Bemba language just fine, yet saw no reason why she should bother to speak in English with her parents, who didn't understand Bemba.

5.4.1.4 The Water's Edge

Don't assume that walking along the bank of a river in northern Nigeria is like walking beside, say, a European river. The water's edge in Africa is dangerous, and I'm still learning that, realizing in 2014 that I was walking too close to the bank of the river in Argungu, where crocodiles lurk. Before the famous annual fishing festival there, I was told that incantations are undertaken to pacify the resident crocodiles, in the belief that this will keep the fishermen safe. In Gaberone, Botswana in 2006, a cobra raised its head and hissed as I walked beside a lake. I was able to back off, but with a crocodile there is no warning, as it can lurch both horizontally and vertically out of the water to snatch a fisherman on the bank, a canoe poleman or his passenger in the Okavango delta, or a bird perched on a dead inundated tree trunk. The day I write this (September 15, 2017), a British journalist was reported by the BBC to have been taken by a crocodile from a Sri Lankan riverbank where he had gone to wash his fingers. Many others die like this, deaths which never reach the international media.

5.4.1.5 Kyrgyzstan

The lady selling me fruit in Kerben bazaar (Jalal-Abad region) on October 29, 2015, has not always worked there. Before the collapse of the Soviet Union in 1991 (and the economies of its dependent States), she used to be a medical orderly in the government health system, leaving its employ when she could no longer support her family on her meager salary there; she could make more money selling fruit and vegetables. Similarly, the onion farmer I met in the field in Shamaldysai village, Nooken raion (district), on the same day, has not always farmed onions—he used to be a medical doctor, but was unable to take care of his family sufficiently on a doctor's salary in a public hospital after 1991 in this "transition" country, in the face of free market price inflation.

5.4.1.6 Yemen

When I was evaluating an irrigation project in Hoddeidah in Yemen, I was told that there was equitable distribution of water from the dam across the spate canal to all members of the "user group". This latter included a VIP figure, having support from the President of the day. Indeed, I could see that the level of water in that person's concrete canal was the same as that for all other members on the day of my visit. Yet I also noted a very healthy horizontal line of algae high up on the canal wall. Clearly we evaluators were being misled. On days when we were not there, the VIP had far more water in his canal than others, by differential opening of the sluice gates. This member of the user group was clearly more equal than the others. Anecdotal evidence must be tempered with critical curiosity and observation to secure the truth.

5.4.1.7 Surprising Attitudes of the "Lightly Traveled"

I was visiting a consulting company based in Rome once, and in a social break the subject of a rock band called Genesis was raised by my Italian colleague Barbara. That I'd never heard of the group elicited a yelp of disbelief. "It's not possible!". "Yes, it is possible Barbara. I spent my youth in Uganda in the 1970s, dancing to Congolese music that was the rage there. I didn't know of the Genesis rock group or its music then, any more than you have heard of the Congolese stars of those days, Tabu Ley and Franco".

When I first left United Kingdom to work in Africa, my assumption was that pineapples grew on trees, like mangoes. How wrong I was. When I worked in Japan in 2005, I showed some photographic slides of my years in Uganda to my mature students. Their incredulous responses amazed me. For instance, a slide of the beautiful campus of Makerere University elicited: "You have universities in Africa?" and "You have trees in Africa?". Yes, and many people in Africa are better educated that many people in Japan, and live in better houses.

I heard a visitor from Africa on her first visit abroad, to the United Kingdom, express huge surprise at the green fields she witnessed, for she had heard that Britain was "industrialized". Her young relative who was with her, on seeing the forests in the county of Berkshire, excitedly exclaimed: "The jungle!". Certainly there was no jungle in Kampala where he lived. Another African visitor was aghast at the skyline of Dubai as the plane came in to land, for she had expected to see only sand dunes and camels. And an American visiting Israel for the first time was heard to say to her husband as the plane came in low over Tel Aviv, before dawn "Hey Herb—they got power!".

5.5 INTERACTION WITH LOCAL ADMINISTRATIONS

It should not be assumed that all in a community welcome help from outside—local government administration officials may be used to charging villagers informally for their help, and won't take kindly to someone setting the precedent of helping free of charge. In addition, direct or indirect approaches may be made by local officials to secure personal benefit from funds being disbursed, and they may be implicated in rigging bids for tenders related to the project. I speak from experience of all these. Unfortunately, the concepts of fairness, honesty and integrity are often in short supply. The official rationale for such behavior is that public sector salaries are too low to constitute a living wage, and government leaders go along with the corruption as they cannot afford to raise those salaries.

5.6 LACK OF TRUST WITHIN MULTIETHNIC NATIONAL COMMUNITIES

Lack of trust among different ethnicities within a country makes community development an especial challenge, as working together to achieve shared benefit becomes far harder if not impossible. Families become isolated, restricted to their subsistence agriculture or livestock pursuits. Cooperatives, with their potential economies of scale cannot start let alone prosper, and the idea of having a communal store for a raw agricultural commodity to await the rise in market prices is impossible to conceive. Similarly, in theory a communal grain reserve may be a great idea to build resilience, but it cannot be applied until attitudes change and trust is rebuilt. The author has experience of such in Liberia, Myanmar and Kyrgyzstan, and most recently in The Gambia—where parents would not work together as a group to raise funds for a particular school, as the efforts of the majority would also benefit parents and children of a particular ethnicity which, by association, was linked to discriminatory actions of the former Jammeh administration. I'm told by Gambians that "non-participation" is the social and institutional culture, and development workers have to negotiate with this norm.

5.7 CONFLICT- OR POLITICAL-BREAK WITH TRADITION

People in countries just emerging from conflict have, of course, a mindset quite different from those who have been enjoying peace and stability for a while. For example, Liberian children don't know about their past, says James Jallah, long-time Yekepa resident and cultural heritage assistant/camp coordinator of mining company ArcelorMittal's community liaison department, in northern Liberia. It was clear during my meeting with him in 2011 that he has the institutional memory for the area, preceding the civil/ethnic conflict (1989–96 and 1999–2003). Because of that tribal conflict in two separate "civil wars", the children were always on the move, he explained to me. Without trust between members of a community there can be no coalescing of individuals around common goals. Without a stable history, it has proven difficult to build a future.

In 2011 also, I was being driven along a remote dirt road in a 4WD vehicle in central Liberia, and in the distance saw a group of people coming towards us on the road. Instead of slowing down to ensure safe passage through the group, as I would have done had I been the driver, our driver accelerated. The group scattered to the edge of the road to allow the vehicle to pass at speed. Once we had gone through, the driver explained to me why he had not slowed down. I could see that he was agitated. In the vehicle we had all seen from afar that the pedestrians were carrying long items. The driver had assumed that the individuals constituted a militia group carrying guns or staves, and the situation was life-threatening for us—it was "them or us", as it had been until 2003. His mindset was still there—the country was still at war with itself. He had not moved on. Even when we were almost upon the group, and he could see that what each was carrying was a shovel (for the group was repairing the road), he did not slow. My two Liberian colleagues in the vehicle, from the Ministry of Planning and Economy, understood the explanation well enough, whereas I was still on a learning curve. Even just days before having seen the destruction wreaked by those militia during the civil war—a lot of heavy machinery at an opencast iron ore mining site in the north having been incinerated, and the western coastal town of Robertsport raised and ransacked—I had not realized how that experience was still raw and present for Liberians, almost a decade after the conflict had stopped.

Another example, from Uganda in 1972, is when I was sitting in on a lecture to second year undergraduates at the university farm near Kampala. Outside the lecture theatre, the jib of a mobile crane touched an overhead electric cable, and there was a sudden loud "crack". All the male students jumped out of the windows and dispersed. It took me a few minutes to understand the reason for this, but I came to realize that it was because we were living under the rule of General Idi Amin, and his reign of terror was ratcheting up. The students had assumed that the campus was being raided by his soldiers, and headed for the relative safety of the banana plantation.

Both these examples from developing countries show that not everyone is ready to engage in long-term development immediately, and the assumption that "development" can happen at the touch of a button may be misplaced. The security and social context is of over-riding importance for the people who live there. In Europe too, the Nazis destroyed Warsaw, but when the Soviets took control after the Second World War, their planners facilitated local architects to rebuild in local style, simulating that which had been wrecked by war, knowing that otherwise the Poles would always be a lost nation.

Sometimes the "conflict" can be between political systems. When one changes suddenly to another, it can render such "transition" societies confused and unstable. Such came about in Soviet satellite countries following the collapse of the Soviet era in 1991. In Kyrgyzstan, for example, people on the one hand are glad to be "free" and with a "democratic" government, yet they also hanker after the stability and surety of the old Soviet system, when pensions were sufficient to ensure a fair standard of living because the government subsidized the prices of coal and foodstuffs, ensured markets for the silk and other agricultural industries, maintained the irrigation systems on State Farms, and so on. Yet providing such social benefits helped to bankrupt the Soviet system, causing its collapse.

5.8 MANAGING EXPECTATIONS

When a donor project sets up in a community, there are often unrealistic expectations. This is an extension of a one-to-one scenario, in which largesse from an outsider in paying a child's school fees expands into an expectation that other physical and financial needs will also be met by the individual philanthropist. The concept of personal sacrifice to enable that person to provide a gift of any sort is often absent. In a rural context, the local perception of a foreigner, or indeed a national visitor from the capital to a rural location, is that the individual is loaded with money and really will not miss it or gift in kind that is sought or provided.

Individuals in some countries are particularly "expectant", with Liberia at the top of my personal list. The reasons are probably multiple, including that the country's nationals have become aid-dependent, and that their self-esteem and confidence to make a difference to their own lives through personal initiative is very low due to recent conflict and its engrained brutality. Personal attitude is also an issue, whereby the concept of "earning" the gift by providing a service is simply not conceptualized.

So, unrealistic expectations of the participating community need to be managed upfront, otherwise even before it starts, the project is set on a course to be branded a failure. Yet this can be counter-balanced by emphasizing that there is much to be achieved if all participants are determined to excel. In 2011 during a mining conference in Monrovia, I heard a VIP Liberian lamenting that in Liberia these days, "Mediocrity is the new norm". This was a devastating appraisal of his country that must be fought in an endeavor to raise the individual and national game. Chapter 10 is devoted to the role of champions, and we can all be champions.

REFERENCES

Laming, D., 2004. Understanding Human Motivation: What Makes People Tick? Blackwell Publishing, Oxford. 311 pp.
Mintz, S.W., 1985. Sweetness and Power: The Place of Sugar in Modern History. Penguin, New York.
Nanda, S., Warms, R.L., 1998. Cultural Anthropology, sixth ed. West/Wadsworth, Belmont, CA.
Scammel, G.V., 1989. The First Imperial Age. HarperCollins Academic, London.
Wilkie, D.S., 1988. Hunters and farmers of the African forest. In: Denslow, J.S., Padoch, C. (Eds.), People of the Tropical Rain Forest. University of California Press, Berkeley, CA, pp. 111–126.

CHAPTER 6

Identifying and Prioritizing the Challenges Confronting Food Security Resilience for All

Contents

6.1	The Need for Resilient Food Systems	72
6.2	Better Policy Making and Planning	76
	6.2.1 Introduction	76
	6.2.2 The Tamil Nadu Integrated Nutrition Project (TINP)	81
	6.2.3 East Africa	83
	6.2.4 Sudan	84
	6.2.5 The World Food Program in The Gambia, Peru and the Gaza Strip	84
6.3	Education to Build Resilience	91
6.4	The Peace Dividend	93
	6.4.1 Absence of Peace	94
	6.4.2 Absence of War	98
6.5	Other Priorities to Improve Resilience	104
References		109

If so many people and so much funding are directed at trying to resolve the problem of food and nutrition insecurity, why is it still nowhere near being fully resolved? Could it be that we are not addressing things as they should be addressed, through paying more attention to *how* we do, rather than *what* we do, for example? Is the nub of the conundrum perhaps that we are still not sufficiently focused on building sustainable structures in society which confer resilience, and preventing or mitigating conflicts, poverty and natural disasters?

Ideally, the community or government must have a vision of what needs to be done, in a prioritized manner, a strategy and action plan to implement it, with funding and *pro bono* contributions in place. Good management, timeliness, monitoring and feedback on the same, and implementing remedial measures, are essential. Above all, the need to build resilience of individuals, in the community and in the government, is paramount to secure sustainability.

Human Resilience Against Food Insecurity
https://doi.org/10.1016/B978-0-12-811052-2.00006-8

Some of the many suitable candidates for prioritization are discussed below, relating to the decisions needed, and who makes those decisions.

6.1 THE NEED FOR RESILIENT FOOD SYSTEMS

A prime example of an overarching challenge to global food security is that the high input *food systems* currently operating in the developed world are unsustainable. This is true with regards to the production system, marketing, distribution and consumption of that food. The term "food systems" used here refers to the generic means of producing food at a global level, and ensuring that it is distributed to those who need it, at an affordable price and in a recurrent manner. This applies not just to the developing world; for example, from a survey of some 41,000 households in the United States, 12.3% (15.6 million households) were food-insecure at least sometime during 2016, including 4.9% with very low food security (Coleman-Jensen et al., 2017).

For a window on "unsustainable food systems" in developing countries, climate change and increasing drought periods, combined with wind erosion, have taken a heavy toll on livestock, as illustrated in Photo 6.1. The huge area of dehydrated carcasses of cattle and draft animals was encountered by the author in 2016 outside of Hamish Koreib village in the northeastern hyper-arid part of Kassala State, Sudan. The animals had died

Photo 6.1 Livestock starvation graveyard, Hamish Koreib, Kassala State, Sudan (Friday April 28, 2016).

not of thirst, as a solar-powered water pump was installed there in 2013 and there are also wells, with water available *ad libitum*. The "graveyard" was right next to that pump and drinking troughs. It was drought-related lack of fodder, from the bush and meager rainfed cereal stover, which had led to their mass starvation.

As for developed countries, a window on "unsustainable food systems" is provided by the statistic that it takes 17 kg grain to produce 1 kg beef under intensive feeding, a highly inefficient use of land. Can the planet afford such a high "ecological hoofprint" (Weis, 2013)? Those promoting the "Meatless Mondays" movement believe that it can't, this initiative having been started in the United States in 2003 by Sid Lerner in association with the Johns Hopkins Bloomberg School of Public Health. There are countless other groups in developed countries lobbying for a lower consumption of industrially-produced meat on "unsustainability" and moral grounds.

Furthermore, as discussed in the earlier book, much of the world's arable land surface is devoted to crops destined for biofuel. Are the fuel needs of the rich more important than the food needs of the poor? Commentators have increasingly spoken in recent years about the originally-Marxist concept of "immizeration of the peasantry" at the hands of corporate capitalism, of increased rural instability, and of ecological degradation in developing countries. "Food politics" is unavoidable.

More resilient, sustainable and equitable systems need to be identified. Some "alternatives" such as organic or Fairtrade endeavor to address how we can better husband our soils, eat healthier food, and provide fairer remuneration for the food we buy from poor farmers. Those who follow these routes may feel that they occupy the higher moral ground. Yet what about those who cannot afford to pay the higher prices for organic and Fairtrade products? Not all farmers can be organic either, as most farmers in developed countries have high fixed costs which could not be met by the lower yields of crops grown organically. So is it only rich people who can live ethically? There is a lot of incipient anger in the existing food system and agriculture in developed countries—in the United Kingdom for instance, farmers have been demoted to employees at the behest of a few large and wealthy corporations, which make the important decisions on which crops are grown, the price paid for them, and the inputs needed to grow them. We should be angry, and politicize, rather than saying we can do nothing. And we *are*, with more and more people becoming involved in making agricultural decisions.

J.K. Galbraith in his book *The Affluent Society*, first published in 1958, pointed out that using Gross Domestic Product as a measure of society's wellbeing takes no account of social and personal indicators of wellbeing. While creating employment, the "must-have" rationale for living leads to overconsumption and waste of natural resources, and encourages youngsters to leave school at the first opportunity to invest in their present rather than in their future. Global food waste could feed 2–3 million people.

I concur with political scientist Susan George, who for decades has maintained that our research efforts need to focus less on poverty and more on affluent throw-away nations (George, 1976, 2004). She argues that these have imposed a nearly universal economic system on the rest of the planet. We, as individuals, institutions and societies, need to search for ways for those of us in rich societies to want less. George says we need to study affluence, and the ways we use food.[18] More people these days suffer from eating too much food, compared with the number eating too little. This is hardly something of which the affluent society can be proud.

The food system that has evolved over the past half-century is designed for a single purpose: maximum efficient production for short-term economic return. To achieve that goal, we have created farming systems that feature increased productivity per unit of labor and land, specialization and economies of scale. Furthermore, our public policies for agriculture have been designed to further that singular goal. But there is nothing in that goal, or the policies needed to support it, which is designed to achieve sustainability—a resilient food and agricultural system (Kirschenmann, 2013). The cited work explores alternative food and agricultural systems that foster public policies and research agendas, to produce resilient food systems that can maintain productivity in the face of future challenges—the end of cheap energy, depleted minerals/fresh water/biodiversity and more unstable climates.

Kirschenmann opines that in our agricultural production techniques, we need to shift from a one-dimensional single-tactic strategy to a systems-restructuring strategy as the guiding principle of our agricultural research agenda. This would lead to the development of systems that focus on harnessing inherent strengths within ecosystems, their "ecological capital". The single-tactic approach is costly in terms of economics and the environment—loss of soil structure due to mining of organic matter and the need for more irrigation water, depleting aquifers, rivers and lakes while increasing soil salinity. Kirschenmann gives examples of farming in the United States, in which while gross farm income increased over time, net farm income has remained fairly

[18]The original title for Galbraith's *The Affluent Society* had been *Why the Poor are Poor*, a study of poverty, but was changed at the suggestion of Galbraith's wife.

constant. This is the farm economy story over recent decades, with farmers being left out of the enormous growth in value of what they produce and sell.

Research should in future be refocused on better understanding the synergies and synchronies of the diverse species in each agricultural watershed, to determine how they can be deployed to increase agricultural productivity sustainably, while simultaneously reducing farming costs and enhancing the capacity of the land to renew itself—namely a more sensitive means of sustaining ecological balance and production, as practiced by previous generations of farmers (though "uneducated" and before the term "ecology" had been invented). They were expert in using rotation of crops and pasture in a mixed farming system, using sheep to control chickweed in wheat, farmyard manure to fertilize the soil and so on. Such sustainable methods are often used in developing countries, such as "night parking" in the Sahel and the "moving barn" technique in the hills of eastern Nepal (Ashley, 1999).

In both the developed and developing world, agriculture's role in the food economy is not as important as it used to be. Take West Africa, for instance: today, 40% of the agro-food sector's value added is no longer produced by agriculture (Allen, 2015). Though, of course, agriculture remains a pillar of economies in the region, the food chain's downstream segments are evolving in line with changes in society. Food and nutrition issues are no longer solely agricultural in nature, and agricultural policy no longer addresses them all.

In West Africa today, just as many people depend on non-agricultural activities for their livelihoods as are engaged in agriculture. This is the major transformation of the past 60 years, linked to the population explosion in towns and cities. Today there are 2,000 towns with over 10,000 inhabitants; in 1950, there were only 150 such towns. There are now 150 million urban dwellers, 30 times more than in 1950. Between 2000 and 2015 alone, the West African urban population grew by over 60 million people. This growth is no longer fueled only by rural migration: most of these people were born in cities. As a result of urbanization and income growth, the West African diet is changing. This in turn impacts food security and nutrition. Diets are diversifying, especially in urban areas. More fruits and vegetables and more processed foods are being consumed, with the latter now representing at least 39% of urban households' food budgets. Even more surprisingly, the poorest rural households devote 35% of their budget to processed foods, showing that these are not limited to the urban middle class.

Yet what does an ideal socially-just "food system" look like, from both the demand-side and supply-side? As well as high-octane individual thinkers and social activists, there are many multidisciplinary centers addressing this issue. One is the Cambridge Global Security Initiative, based at Cambridge University. It comprises a network of researchers (crop scientists, economists, engineers, social geographers, policy and public health experts) addressing sustainable food production, supply/processing and distribution chains, and healthy ethical consumption.

The subject of "food systems" is too large to be encompassed satisfactorily in this section without unbalancing the chapter, so the reader is kindly referred to Section 11.1, Case Study 1, in which technical and anthropological factors conditioning the search for improvements are discussed, involving decisions made at numerous levels determining which direction is taken.

6.2 BETTER POLICY MAKING AND PLANNING

6.2.1 Introduction

For a long time, international planners considered it not worthwhile to invest in improving subsistence agricultural systems. However, that perception has changed as a result of evidence-based research showing that investment in such is one of the wisest investments possible, in terms of value for money. This has occurred not only because of this economic rationale though, but also for the social benefits that can result, instilling resilience into marginalized communities and reducing the pressure on youth to migrate to cities to seek work.

Global food security is compromised by insufficient attention being given by planners and policy makers to the complexity of the lives of those who are food-insecure in the developing world. Food availability, access and use depend on a multitude of factors, which are time- or location-specific (Pottier, 1999a). Moreover, what farmers in a community grow, what people eat, how they trade, and their social support lineage and personal networks are linked issues, with each susceptible to external influences. "Projects" so often fail to engage with these cultural, economic and social relationships in their design and implementation. Commonly, for example, in a laudable attempt to inculcate improved rights for women, a project may be "women-focused" rather than being "gender-sensitive", with the result that the project is perceived by men as a threat, as they feel left out. Women can then become even more marginalized in a patriarchal society, the project intervention thereby being counter-productive.

The do-no-harm imperative is so important, yet often insufficiently understood by policy makers. This can result from the lack of political will and hence appropriate policy, such as that exhibited by the then colonial rulers of India during the Second World War in not diverting rice from being channeled to the war effort for use in India itself, which was perhaps the major cause of the Great Bengal Famine of 1943–44 (Ó Gráda, 2015); this contests the view of Sen that it was due to speculation and market failure (Sen, 1981). Policy failure often relates to donor interventions too, in situations of fragility and conflict. Donors or foreign governments can inadvertently do harm when the resources they deliver or the policy reforms they advocate exacerbate rather than mitigate the conditions for violent conflict, or they weaken rather than strengthen the State as a result of decision making and policy formation over the deployment of public resources (Box 6.1).[19]

BOX 6.1 A Do-No-Harm Failure in the Philippines in the 1960s

An article by Bruce Nussbaum on December 3, 2014 illustrates how technical advances and "peace" are intertwined, and how an apparent "advance" can lead to the suffering of marginalized peoples.[20] Nussbaum was a Peace Corps volunteer in the Philippines in the late 1960s, and participated in a trial to establish whether newly-introduced Japanese tractors were more efficient than the traditional water buffalo during cultivation of the newly-introduced high-yielding IR 8 "Miracle Rice", developed by the International Rice Research Institute (IRRI). The setting for the trial was the mountains of Pampanga and Tarlac, in the region of central Luzon. At that time, a collection of oligarchic families dominated the Philippines' political system and owned much of the land, which they ran in a feudal manner.

The Japanese tractors proved more efficient, yet just as for Miracle Rice, their introduction had then-unforeseen outcomes. Though IR8 could treble the yield of the local types in Luzon, to do so it needed far more inputs (fertilizer, insecticide and water). Moreover, there were clearly additional costs to procure and operate the tractors. These incremental costs were unaffordable to peasant farmers, whereas the powerful families could cover the additional financing. There were social costs too, for the small farmers could not compete against the oligarchs. The promised government loans for peasants were largely unfulfilled, and their water buffalo depreciated in value, as their use had been largely usurped by the tractor. Unable to earn sufficient money by selling their buffalo or to compete against the new technology, the small farmers ended up losing their land to the oligarchs.

Rather than representing a boon, the coming of Miracle Rice and its associated technological package and economies of scale put an end to the livelihoods of the independent peasant class. Their disenfranchisement inflamed social tensions. Luzon has a history of rebellion of the masses against feudal landlords; it was the focal point of the Hukbalahap insurgency of 1946–54, for example, which almost toppled the government of the day. The introduction of Miracle Rice resulted in central Luzon once again erupting into conflict, involving a proxy war between major world powers which backed the insurgents or government.

---

[19]https://www.oecd.org/dac/conflict-fragility-resilience/docs/do%20no%20harm.pdf (accessed September 22, 2017).
[20]https://www.moma.org/interactives/exhibitions/2013/designandviolence/miracle-rice-international-rice-research-institute (accessed August 16, 2017).

Using evidence from the research of social anthropologists across a range of cultural contexts in the developing world, Pottier (1999a) considers "modern" food production, marketing and distribution systems which are having a strong impact on the livelihoods of subsistence farmers, who are vulnerable to food insecurity and famine. In chapters relating to agriculture, marketing and famine management, he draws attention to the contrast and lack of congruence between the realities and perceptions of those who carry the burden of food unavailability or unaffordability on their backs and those of policy makers, multinational food companies, donors, governments and NGOs. The powerful international grain lobby, for example, has little interest in the nutrition challenges faced by poor villagers or slum dwellers.

Applied anthropology can trace the dynamic linkages between the local decision-making unit and larger society, by charting the gulf between real and assumed behavior and by paying attention to past experiences, all of which are part of their ethnographic orientation. By translating the needs of local populations and adopting a client-driven goal structure within broader policy contexts, anthropologists can nudge the formulation of a public policy from one that will likely fail to one that will likely succeed (Nanda and Warms, 1998).

Anthropologists have long worked on public policy issues, in an applied and largely uncritical way, in the service of governments, a tool of governance and power. More recently, this has been deemed an "undersell", a blind spot for anthropologists, for "policy" itself deserves to be subject to anthropological enquiry (Shore, 2012). The potential is there for such enquiry to inform and guide government (or other bodies) in formulating "good" policies. The modern approach to anthropological engagement with policy recognizes that policy has become more and more a dominant organizing principle of contemporary society, which orders and regulates its behavior, whether these policies derive from government, public institutions, NGOs or the commercial private sector.

Development workers in the field need to interact with these institutional decision makers, both large and small, and seek to influence them and work with the decisions they make. Sridhar (2008a) uses a number of case studies in health and education related to South Asia to consider the process whereby anthropologists endeavor to understand how organizations work, and their interaction with their clients. Throughout her volume, the methodological and practical issues involved are highlighted, including "the language barrier, the unreturned phone calls, the tense relationships and the unhappy managers", and the coping strategy that anthropologists need to fashion. Prior to this monograph there were other published works addressing "organizational anthropology", which though largely relating to the developed world, nevertheless clarified why anthropologists needed to address policy issues at all (Wright, 1994; Shore and Wright, 1997), and the objectives that should condition such anthropological enquiry (Bate, 1997; Gellner and Hirsch, 2001).

Anthropologists are now exploring ways to increase the impact of the discipline on public policy, especially development policy. Sponsors of the panel "Conflict and Compromise: Perspectives on the Practice of Anthropology" at the 1987 American Anthropological Association meetings lamented that despite 25 years of practice, development anthropology had by then had minimal impact on policy. Cases in which anthropologists *have* had a significant impact on development policy in West Africa until that date have been documented (Arnould, 1989). A fundamental problem constraining leverage which anthropologists had on policy makers until 1987 was that development policy had been formulated under a paradigm which assumed that developing world populations should adapt to the conditions of "economic reproduction" imposed by capitalism: namely, the physical production, distribution, trade and consumption of goods and services. Anthropologists who accept this paradigm compromise scientific standards and their professional ethics. They need to maintain their integrity when faced with any pressure from funding agencies to distort or suppress their findings and recommendations when they come into conflict with agency agendas.[21]

Since those 1987 meetings, anthropologists have made important contributions to public policy formation by providing data on the groups toward which policies of change are directed. An important area of applied anthropology is in programs of planned change or development in village and hunter-gatherer communities. Development anthropology increasingly identifies the link between local communities and larger global contexts.

Ferguson (2006) counseled that more attention be paid to the formidable institutions that are "governing" Africa from afar, especially the transnational financial institutions, such as the World Bank and International Monetary Fund (IMF) and development agencies (UN agencies and donor organizations). These powerful, largely unaccountable transnational institutions had

---

[21]Not only anthropologists should maintain their integrity like this.

been studied very little, yet play a central role in the *de facto* governance and administration of the continent. The decisions of their unelected officials affect much of African society and its economy. The concept of the sovereign African nation–state is a myth, Ferguson convincingly argues.

Major social and economic indicators for African countries began to decline after the 1970s, which led to structural adjustment programs after the 1980s, which promoted the private sector at the expense of the public sector. This was intended to address food security and nutrition security only indirectly, through improved economic stability. Later, the World Bank and donor community modified this approach to be more balanced, with the Cooperative Development Framework and Poverty Reduction strategy paper. Once the New Partnership for Africa's Development (NEPAD) was initiated in 2001, African initiatives and ownership of the development process could be integrated with neoliberal concepts. In much of Africa, the private sector is still weak, and progress is further constrained by an uneven playing field in international trade and hemorrhage caused by conflict and war (Heidhues et al., 2004). Agriculture needs to be at the forefront of the agenda, with relevant policies designed which address consistent objectives, and these policies should be implemented under decentralized planning.

Anthropological research is often overlooked when policy is made, policy makers being confronted by partisanship in a political party (government) which is trying to please its financial backers and industrial/multinational lobby groups, for example. Applying anthropological knowledge in policy making may extend beyond a shortfall in cultural awareness in development agencies, and move into the orbit of politics. Simply publishing research findings is quite inadequate for anthropologists to change public policy. Policy formulation is a political process and does not proceed on the basis of available scientific evidence alone. To create their own lobbying, anthropologists need to "take the gloves off".

As indicated above, an overarching challenge is that there is often a disconnect between those in development organizations or the public sector determining agricultural, trade or nutrition policy and those who are most affected by implementation of those policies. An example of banana production in Belize was given as Case Study 3 in the companion website to the earlier book, in which policy is driven by industrial private sector interests. Another case of such incompatible and contrasting perspectives is given by Sridhar, between the poor people of the target group in southern India and the World Bank (see Section 6.2.2).

6.2.2 The Tamil Nadu Integrated Nutrition Project (TINP)

Organizational anthropologist Devi Sridhar (2008b) examines the role of the World Bank in a nutrition project in India, and the reasons why hunger and undernutrition have not been cracked there. How did a nation that prides itself on its agricultural self-sufficiency and booming economic growth come to have half of its pre-school population undernourished? The Indian government spends more on grain reserves than on agriculture, rural development, irrigation and flood control combined (Waldman, 2002). Why have strategies to combat hunger and undernutrition failed so badly?, she asks. Her book addresses these questions and probes the issues surrounding development assistance and power structures, hunger and strategies to eliminate undernutrition, and how hunger should be fundamentally understood and tackled.

Throughout the book, the underlying tension between "choice" and "circumstance" (external factors—societal, economic and political that influence the choices that individuals make), is explored. To what degree are individuals able to determine their life choices, and what are the external constraints which determine behavior? Sridhar's book is not just about nutrition, but an attempt to uncover the workings of power through a close look at the organizational structure of a department in the World Bank, how it defines and tackles problems it seeks to address, and how policies are formulated. In this process, the source of nutrition policy in the Bank is traced to those affected by those policies in India. The book frames hunger and nutrition as societal issues, involving the challenges of poverty, colonial heritage, land deforestation and gender inequality, with undernutrition viewed as an expression of community sickness.

This evaluation study by Professor Sridhar, pursued through observations and interviews, both in Tamil Nadu and in the World Bank, takes the form of investigating the anthropology of the organization in relation to its work in southern India, with the purpose of trying to understand "what works to effectively address hunger". She draws attention to both the successes and failures of a particular World Bank-funded project in India—the long-running Tamil Nadu Integrated Nutrition Project (TINP), which the Bank had used as a template to design nutrition projects in other parts of the world. This project was a growth monitoring, food supplementation and intensive nutrition counseling program, with services provided at a community nutrition center. The TINP was regarded as a success by the Bank, following the 1986 mid-term evaluation (MTE), which recorded a reduction in rates of severe undernutrition.

The MTE also deemed that the TINP was very cost-effective, the cost per death averted being estimated at $US1,482 per year, much lower than other mortality prevention projects such as malaria control and polio immunization (Berg, 1987).[22]

The key failure of the TINP, Sridhar argues is the erroneous central tenet of the Bank and the TINP relating to nutrition, namely that undernutrition is the result of inappropriate childcare practices rather than the result of low income or other "circumstance". A subsidiary failure was the silo mentality within the Bank structure, with various departments doing their own thing more in isolation than in concert, so that poverty and gender transformation imperatives, for instance, were not (though could have been) incorporated in project design. The Bank was less than forthcoming on investigating the shortcomings that were pointed out, and Sridhar is critical of the structural ideological and political constraining factors on decisions that Bank staff made, though she developed respect for many individuals working within the Bank. The Bank should be amenable to challenge, despite its prestige and power; this is so important for several reasons, not least that for the TINP, for example, Bank funding was not a grant but a loan, which needs repayment.

The objective of the project was to improve nutritional status, and it was planned and implemented by nutritionists. However, the multisectoral causes of the poor performance for nutritional indicators in the target community were not addressed in a coherent way. Despite the Bank having policies on mitigating poverty and gender inequity, for instance, the practical means to reduce the occurrence of these in order to improve nutrition were not marshaled during the TINP implementation. Rather, the decision-making processes of families were addressed in relation to their coping strategy to overcome circumstances, via choices they made—for instance, through rearranging disbursement of disposable income across various categories of food.

It would have been better had the project, and the Bank, focused less on interventions such as fostering policies that provided health education and nutritional knowledge, from which individuals could choose different behavior such that better nutrition was fostered, and instead prioritized addressing the contextual circumstances which leads to undernutrition in the first place: namely, inadequate purchasing power and gender inequality,

---

[22]Though various commentators identified shortcomings in the methodology used during the MTE and subsequent evaluations.

the main drivers of undernutrition in India. These two aspects of actual circumstance could have been targeted through creating more opportunities for people to secure or enhance their incomes, enabling improved access to food, and more nutritious food in particular.

Structural determinants of hunger, such as the availability of work and the capacity of people to do it, and the pervading patriarchal norms which heap social and economic disadvantages on women, the main caregivers of a community's children, would have been a better focus of the TINP, suggests Sridhar. Chapter 3 and Chapter 4 in her book, concerning the understanding and addressing of hunger within the World Bank, are required reading for those interested in food and nutrition security as it can be influenced by a development project funding agency.

In her book, Sridhar argues that there was a mismatch between how undernutrition was defined, measured and evaluated by the Bank, and how it was/is lived and experienced in affected communities. Policy makers have to recognize that undernutrition is not a "disease", but a biological manifestation of a nexus of political, economic and social forces, the interplay of gender, caste, class, household and community dynamics, which cannot be overcome through a program rooted in growth monitoring and nutrition education alone. Nor can the causes be eradicated over a short time frame. As the multiple causes of undernutrition are "structural", so they need redress in a "structural" way. There needs to be a coming together, an improved dynamic, of those who make policy (in development banks and governments especially), front-line implementers of policy in the community, and the final beneficiary community itself. The processes whereby socioeconomic forces become manifest in a measurable biological form need to be understood by all those who seek to improve the *status quo*. And of course, the holistic effort needs to be community-led, with the marginalized in that community thoroughly engaged with the actions undertaken, understanding them and having full empathy with them—working *with* people, not *for* them (see Section 4.3). This will lead to better evidence-based policy making, rather than policy-based evidence making, which Sridhar deems was an unfortunate hallmark of the TINP.

6.2.3 East Africa

Some years ago I was involved in a public dispute between a Ministry of Agriculture, an international bank and an agri-processing entrepreneur on the one hand, and the EU Delegation, the National Department of Forests and the National Environment Agency on the other. The proposal was that the pristine forest on some islands in Lake Victoria should be cut down and replaced with oil palm. I was part of the contesting group, on the grounds that the forest had intrinsic ecological value and could bring in far more revenue for the nation and local communities from discerning tourists than could oil palm, which would be growing at the limit of its profitability at the altitude of Lake Victoria, albeit on the equator. I made personal visits to the development bank headquarters and private sector investor in Nairobi to explain my concerns, and drafted an open letter to the press from the EU Delegation. Those of us in the contesting group were labeled "enemies of development" by the national president of the day. The compromise result was that the project went ahead, but with a much-reduced footprint, which itself must have hit the potential economy of scale benefits for the company.

6.2.4 Sudan

In 2016, I evaluated a food security program in Sudan. I maintained that the hi-tech modern agriculture being undertaken in the vast swathes of the plains was unsustainable. The same had been said by UNEP 9 years earlier (UNEP, 2007). My three Sudanese colleagues and I advocated a Conservation Agriculture approach, in which a soil cover is maintained throughout the year, rather than bare soil surface—if maintaining such a cover were possible with the prevalent customary free grazing that is practiced by the people, this to be determined (see Case Study 2 on Conservation Agriculture in the companion website of my earlier book). The donor did not accept this initially, and even one of the NGO implementers said we were "against modern agriculture". No, we evaluators were in "post-modern agriculture" mode in our recommendation. Dust storms across the bare fields were a common sight on the dry plains during my stay in Sudan, and tellingly on May 17, 2016, 1 mm of topsoil was deposited by the wind overnight across Khartoum city, where I was staying. Dust storms have been a feature of Sudanese life for the last 50 years, according to the FAO office, Khartoum.

6.2.5 The World Food Program in The Gambia, Peru and the Gaza Strip

Three "problem" areas have been cited above, in India, East Africa and Sudan, caused by bad policy and planning. Yet there is a lot of good news on policy out there too. One instance involves the work of the World Food Program (WFP) in its school feeding and other programs. On the face of it, giving free food to schoolchildren, or food vouchers to families, may smack of humanitarian aid, which is unsustainable in the sense that it requires a continual external financial input in perpetuity. Yet neither of these characteristics necessarily holds true, and they can be used to leverage and encourage local enterprise, in the interest of building community resilience. Three examples are given below, all of which involve joined-up governance of the stakeholder groups in the interests of democratic credentials and building resilience of the poor.

6.2.5.1 The Gambia

There is an excellent School Feeding Program in The Gambia. The promise of a meal, a nutritious one, at school is a huge incentive/reward both for the child coming to school, and for the parents sending their child. It may be shocking for someone who has never ventured from prosperous parts of the developed world to understand the socioeconomic condition of poor families in The Gambia, or elsewhere in the developing world. In a world of poverty, the child goes to school without a substantial breakfast, or perhaps none at all. In The Gambia, the walk to school through the bush may take up to 2 hours. On arrival at school the child is in no condition to learn, or even stay awake. The physical and physiological condition of the child is improved enormously as a result of a daily school meal being provided—these are measurable attributes. From collected quantitative data and opinions captured from parents, school staff and pupils, there is a strong positive correlation between the School Feeding Program and pupil enrolment, attendance, retention, punctuality, attentive capacity and performance in class and exams.

What better example of an intervention to build resilience in a community can there be? If an individual has damaged cognitive capacity through undernutrition, has not benefitted from learning at school, and does not pass the school leaving exam, what hope is there that that individual can integrate with society in a productive way and be assured of a rewarding livelihood and socioeconomic condition which surpasses that of his or her parents? Food and nutrition security and livelihood security are positively correlated.

Implementing partner the WFP, excellently led by its Country Director (nutritionist) Angela Cespedes, has managed a school feeding program in The Gambia well, with an equally excellent public sector partner having full buy-in, the Ministry of Basic and Secondary Education (MoBSE), now led by technocrat Minister Claudina Cole.[23] Financial support is forthcoming,

---

[23] For example, the national School Feeding Policy was formulated under the WFP/MoBSE effort of the MDG1c project, and is awaiting endorsement by Cabinet.

currently and for the next few years, from the European Commission, with an excellent manager, Darrell Sexstone, at its Delegation near Banjul; the continuity of funding is essential in the interests of planning. This funding is supported by a ring-fenced counterpart budget within the MoBSE.

This formidable triad of managers, working seamlessly together on the ground, is only too aware that the marginalized communities they have the opportunity and obligation to help, need that help *now*, on *their* timescale. They are ever promoting ways whereby the food items may not only be locally-sourced in support of local farmers, but also grown on school land where that is possible in rural locations. Moreover, the WFP teams in the field endeavor to find ways in which parents, school staff and community members through the Village Development Committees and Multi-Disciplinary Facilitation Teams can between them devise ways to raise money through enterprise to help pay for the school meals, which are currently largely externally-funded. Furthermore, if funds allow, the WFP intends to promote rainwater collection from the tin roofs of school building, and increase the number of taps for washing before meals, mindful of the importance of WaSH parameters in the interests of food security, and the ever-falling water tables in the Sahelian environment which pervades much of The Gambia. These water-related improvements are nutrition-sensitive, fostering improved hygiene and therefore retention of the nutritional value of the consumed food. School kitchen improvements are being fostered too (funds permitting); more efficient wood-burning stoves are in the interests of conserving firewood and the environment, and render the kitchen a less smoky and hot working environment for the cooks (Photo 6.2). Thereby the WFP transforms itself from its erstwhile status of "humanitarian organization" into a "development organization", with sustainability and resilience credentials.

During a working visit to the country by the author in 2017, a particularly good multisectoral farm model was noted at one school in Janak (West Coast Region), organized by the Swedish NGO Future in Our Hands. It was laid out with four components: irrigated vegetables, fruit tree and horticultural nursery, stall-fed small ruminant livestock and bee hives. This is an excellent risk-spreading strategy. It also has the advantage that production and/or income streams from the various ventures combine to provide food for the pupils and/or income with which to purchase food for school meals, for most of the year. The risk-averse strategy and diversified income streams promote resilience. Throughout this school feeding initiative, good government, school, community and project policies and execution are in place.

Photo 6.2 Wasteful use of fuelwood in a school kitchen at Janak, West Coast Region, The Gambia, compared with the preferred enclosed fuel-efficient stove (June 8, 2017).

6.2.5.2 Qali Warma *in Peru*

Peru is an upper middle–income country that has witnessed significant economic gains over recent years. However, many poor people have been bypassed by these improvements, and inequality remains rife across the country, with millions suffering from undernutrition and being highly vulnerable to natural disasters. Their resilience against food insecurity is low. According to the WFP, although chronic child malnutrition fell to 14.5% in 2015, anemia rates have experienced an upward trend since 2011, affecting 46% of children under 3 years. It is estimated that approximately 7.1 million Peruvians, or one in five, live in a district with high or very high vulnerability to food insecurity in the face of natural disasters.[24]

Peru's social safety net comprises five programs, one of which is the provision of school meals through the *Qali Warma* National School Nutrition Program. Like the Productive Safety Net Program (PSNP) in Ethiopia, *Qali Warma* has become a highly respected large-scale program in the field of food and nutrition security. *Qali Warma* means "strong child" in the Amerind Quechua language. It was launched in 2012 and, in terms of budget allocation, is now the largest program within the Ministry of Development and Social Inclusion (MDSI) (OECD, 2016).

[24]https://wfpusa.org/countries/peru (accessed August 3, 2017).

The aim of the program is to address child food insecurity and undernutrition. Its specific objectives are to:

1. provide beneficiaries with quality nutrition throughout the school year, based on their living situation and other factors;
2. improve beneficiaries' attention spans in class, encouraging their attendance and retention; and
3. promote better eating habits among the program's beneficiaries.

Its target group is kindergarten (from 3 years of age) and primary school children in public educational institutions. Through its assistance to *Qali Warma*, the WFP and partners aim to strengthen social protection in Peru by providing technical assistance to the national school meals program focusing on menu revisions, food fortification and the introduction of fresh fruits and vegetables, to address national micronutrient deficiencies. *Qali Warma* delivers milk and other nutritious products to millions of children aged 3–5 across eight regions in Peru, including coastal areas, mountainous regions and the capital city, Lima, home to more than 30% of the population. More than 50,000 kitchens with equipment have been established under the program. It also has a strong educational element, with children being taught about the importance of healthy eating, personal hygiene, recycling and protecting the environment. The MDSI estimated that it would reach 3.5 million children in 2016. The program is due to be extended across the whole country over coming years.

A field-level example is featured in a report from Agenzia Fides on April 20, 2017, recording that in the extreme poverty areas within the south-central region of Apurímac there are 1,334 public school institutions that receive food from *Qali Warma*, where 38,218 pre-school primary school children are offered breakfast and lunch. According to a note sent to Fides by the head of the social program of the Apurímac Territorial Unit, this figure represents 69% of the total public school structures assisted by *Qali Warma* in the region. The report affirms that 22 kinds of products are distributed, including local foods such as quinoa, wheat, bean flour, barley, corn, wheat, rice, oats, sugar, kiwi fruits, milk, soups, dehydrated eggs, preserved fish, dried sheep meat, minced meat, turkey and chicken. Currently, *Qali Warma* supplies food to public schools in the Apurímac every day of the school year.[25]

Through capacity building across all stakeholder groups, other program beneficiaries include all those involved in organized community education, local government and the private sector. The program is delivered via an inclusive governance system that assigns vital roles to local stakeholders. Each

---

[25]http://www.fides.org/en/news/62129#.WZ2C1j6GPiw (accessed August 23, 2017).

school, for example, needs to establish a Purchasing Committee consisting of public officials and parents, tasked with buying foodstuffs. Similarly, a School Feeding Committee, comprising teachers and parents, is responsible for monitoring the quality of food and delivering it to pupils. Participation of, and co-management with, the community is a cornerstone of the program, as is the involvement with other government agencies—the Ministries of Education and Agriculture, and the National Food and Nutrition Center. International partners include the WFP, FAO, World Bank and GIZ. Interviews with national stakeholders have indicated that the program has successfully addressed health issues (downstream of better feeding) and improved children's learning capacities. Just as for The Gambia example cited above, the success of *Qali Warma* may be ascribed to good policies and their joined-up implementation, and full commitment of all stakeholders for the common good (Box 6.2).

BOX 6.2 Participation in *Qali Warma*

As indicated above, participation under *Qali Warma* is of the "appropriate" type, rather than the converse as characterized by Bliss and Neumann (2008). One important finding of their review of literature and their own experiences is that in many development programs (from project level to sector-wide approaches, and Poverty Reduction Strategy processes) funded by bilateral and multilateral agencies, often implemented by NGOs, participation is envisaged and implemented only in a functional and utilitarian way. Participation is used in order to achieve predefined objectives as laid down by the donor in program/project documents.

Bliss and Neumann found that especially at project and program levels, true participation does not happen, because in almost all cases the most important decisions have already been made by the time the project or program is taken by the donor to the people, and "beneficiary participation" cuts in—for example, decisions on priority areas of the interventions, goals, objectives and activities. In those cases where civil society *does* participate in poverty reduction strategy (PRS) processes, serious doubts arise regarding the legitimacy and representativeness of the participating actors. Small development or business NGOs, and also INGOs, dominate the public debate, whereas member associations and advocacy groups are only rarely consulted. While the many important social actors are able to participate in PRS processes in a number of cases, the participation of poor and disadvantaged population groups is rather unusual.

Participation should serve as a tool for the empowerment of the target group, so that through its involvement it secures ownership of the benefits and outcome of the intervention. True participation should comprise a key component of the donor's exit strategy.

As an example, the government revised public procurement procedures so that local food producers could supply school meals for children between the ages of 3 and 6.

6.2.5.3 WFP and the Economic Recovery Program in the Gaza Strip

The WFP and the United Nations Relief and Works Agency (UNRWA) between them are responsible for providing food aid to most of the 2 million Palestinians trapped in the Gaza Strip by the political impasse with Israel. The WFP, laudably, wishes to source locally as many of its food items as possible, buying from smallholders and small agri-processors in the Strip, thereby stimulating the local economy rather than depressing it with imported goods.

The DANIDA-funded Oxfam-implemented program Economic Recovery in the Gaza Strip: Promoting Choice, Resilience, Dignity (ERGS) provides an avenue to attain this goal. This market development project addresses socioeconomic empowerment and livelihood improvement among vulnerable families and youth. It uses economic recovery as the main vehicle, through supporting potential economic growth sectors (agriculture, dairy, ICT), and promising small-scale enterprises. Outputs from these enterprises may be used as items availed to beneficiaries under the aforesaid WFP program. This is already happening on an increasing scale, and is having a positive impact on the profitability and morale of the micro-processers involved, together with engendering an enhanced regard for the WFP as a development partner. ERGS operates through a "making spaces" approach which allows the flourishing of local business aspirations and comparative advantages.

The WFP program cited comprises an humanitarian response to the food and nutritional insecurity in the Gaza Strip occasioned by the ongoing Occupation. It provides an opportunity for qualifying non-refugee residents to procure food through vouchers or cash, while also promoting the local trading environment. Both Oxfam GB and WFP believe that the current somewhat *ad hoc* approach to their partnership warrants a more coordinated and synergistic upgrade. They have agreed that a feasibility study be mounted both to explore the potential for this and to formulate a roadmap whereby this may be achieved, with specific intervention entry points identified. The outputs expected include better linkages between the two programs, a boost for the Gazan economy and improved policies and practices for the two organizations. Terms of Reference (ToRs) were prepared for "Exploration of opportunities for development of partnership linking aid and recovery actors in the Gaza Strip". The ToRs say that the study will "explore how the linkage between ERGS and the Value Based Voucher (VBB) program of WFP, may be rendered more robust and synergistic".

6.3 EDUCATION TO BUILD RESILIENCE

Ofir et al. (2014) campaign for a systemic and transformative change in current Agricultural Education and Training (AET) in sub-Saharan Africa, such that the agriculture sector at all levels becomes better geared to deliver impact to people's lives and improve their resilience to shocks and stresses. The need for such transformation had already been identified (World Bank, 2007; Juma, 2011; Moock, 2011; World Economic Forum, 2013). Ofir et al. stress some bottom-up and top-down features which are needed to impact resilience, pointing out key levers and potential tipping points to secure such change on the ground, with potential interventions structured around such strategic areas to combat any inertia and resistance to change within institutions and vulnerable farming communities. Leadership, capacity, political will and resources are all needed to support renewal and transformation.

"Transformative change" often requires a shift in perspectives, attitudes and/or values of an individual, group, community or institution, which in turn gives rise to changes in behavior or performance. There will be drivers and enablers (catalysts) of change, helping to dispel any disabling constraints. In the case of AET, interventions need to be sought which maximize the potential for transformative change, as individual interventions or in a package, and decisions on how these may best be sequenced. If not properly envisioned and managed, the system under change may collapse. Simultaneous adjustments to the agricultural market may be needed to cater for consequences of change in AET, to ensure that affordable inputs and services can be accessed in a timely manner, downstream of changes in AET. For that to happen, putting smallholders' needs at the center of policies and plans is crucial.

For the whole intervention plan to result in impact and building resilience, the agents of change need to be well-organized, motivated and capable of performing their given tasks well. Training institutions and their staff can sometimes be unresponsive when it comes to change, especially older staff.[26] Ways of doing things can become set in stone, and a silo mentality

[26]In a project formulation assignment in Borno State, Nigeria, in 1991, I began speaking to some senior rural development workers of the value of the Grameen Bank in Bangladesh, to which one senior person in the room dismissively retorted, "We have nothing to learn from Bangladesh"!. I then set about explaining that we should all be prepared to learn from whichever source.

may be entrenched. It is far better that an interdisciplinary modality is followed, which has many advantages to harness synergy and improve impact, with teaching and research based on the hard realities of marginal smallholders and their priorities. I recall being impressed three decades ago by finding that the government agricultural department in the then-called People's Democratic Republic of Yemen (PDRY) was organized not by discipline, but by agricultural commodity.

Moreover, some areas which should feature in the AET syllabus may not be there, such as coherent policy making and agribusiness, even food and nutritional security and smallholder resilience. The farmer needs information, and means to access it, on a present continuous basis, together with skills and physical/financial inputs in order to apply that knowledge to increase productivity and profit on crop and livestock produce sold. Finally, the farmer needs confidence-building and mentoring support. Both researchers and extensionists (government or private sector), working in collaboration, need to deliver such support seamlessly and in perpetuity, in a people-friendly gender-sensitive way, in full empathy with the farmers' existing knowledge and skills, thereby improving livelihoods, food/nutrition security and resilience. This is a tall order. It is the task of training, research and extension institutions to deliver it.

Pottier (1999b) points out that gender responsibilities in food provisioning would benefit from a "clearer focus on male economic uncertainty and impoverishment", and how men may feel threatened by the economic independence of their wives. Moreover, women who have secured space and time for own-account farming often still face the challenge of playing the market well, as this is difficult in terms of their time management. Such gender-related issues must feature in any AET syllabus in the developing world.

The resilience of an individual, society, country, institution or system depends on its plasticity and ability to react to a shock, through adapting or resisting. As Ofir et al. (2014) point out, the starting point for cultivating resilience is to identify the drivers of current vulnerability. Drivers of such embedded in many AET systems in Africa currently include: insufficient priority and resourcing given to agriculture and public sector staff in the sector; overdependence on foreign assistance with their associated agendas; weak linkage between Ministries of Agriculture and Education, and between agricultural production and markets, to render agribusiness more competitive; lack of economies of scale and scope in smallholder communities; and weak propensity for innovative thinking and methods in the public sector.

In my first job, teaching agriculture at Makerere University, Uganda, I asked my class what their parents did. One student responded: "My father doesn't have a job. He's a farmer". That opened a window for me to understand just how much potential is lost because agriculture was considered something people do when they can't get "proper jobs". The apex challenge for those in the AET system is to demonstrate that farming *is* a proper job, that it is a business, and that it can be a *profitable* business and livelihood, the more so as it expands, whereby adherents can both improve their resilience and secure rural development with better provision of services.

6.4 THE PEACE DIVIDEND

As pointed out in the UN Report on the State of Food Security and Nutrition 2017 in its Foreword, of the 815 million hungry people on the planet, 489 million live in countries affected by conflict. That Report in its subtitle and content recognizes the positive correlation between peace and food security. To address food insecurity and undernutrition in conflict-affected situations requires a conflict-sensitive approach that aligns actions relating to humanitarian assistance, long-term development and sustaining peace.

Breaking the links between hunger and conflict must be a priority for food, agriculture, environment and economic development policies (Messer et al., 1998). This requires that the international community focus better on delivering food security in ways that prevent conflict, distributing food aid in ways that do not prolong conflict, and offering assistance in reconstruction.[27] Peace governs the linkage between relief, rehabilitation and development (LRRD). An example involving variety IR 8 in the Philippines was given in Section 6.2.1. An example from Africa on the intertwined nature of potential conflict and threat to food security is exemplified by competition among riparian nations over rights to use water from the River Nile.[28] Conflict nearly always impinges on the production/availability and distribution of food, and leads to increased competition for food, land and resources. Controlling the production of, and access to, food can also be used as a weapon by protagonists in conflict. Collinson and Macbeth (2014)

---

[27]I draw attention to Master's courses at the University of York (UK) in reconstruction after conflict, highly spoken of by my national colleagues in the Ministry of Rural Rehabilitation and Development (MRRD) (Afghanistan) who graduated from them.

[28]http://www.bbc.co.uk/news/resources/idt-sh/death_of_the_nile (accessed October 1, 2017).

examine the roles of specialists in anthropology, nutrition, political science, development studies and international relations, in helping to understand linkages between hunger and conflict, and consider the perspectives of practitioners working in the private and public sectors who engage with food-related issues in the field.

This section discusses the "peace dividend" that, though often elusive, is a necessary albeit on-its-own insufficient factor for building resilience in a community, through creating an enabling environment. The more restricted meaning of the term is the "economic benefit anticipated as a result of money previously spent on defence/conflict becoming available for other purposes".

6.4.1 Absence of Peace

Clearly, unless peace prevails in a country there will be precious little community or national development there. Without peace, food and nutrition security in the conflicted area will become more endangered, as its inhabitants need support through humanitarian assistance. Global demands for this assistance are increasing and far surpass its availability, as it has to be shared with victims of natural disasters (these are increasing too, linked with global warming). Countries currently at war within themselves and/or with neighboring States are not doing well, even moving backwards, for example, Libya and Yemen. Syria and South Sudan are in ruins. Myanmar is marking time as the peace process recedes, and Burundi is in suspended animation owing to the political impasse at the top. Palestine continues to de-develop in the face of its continued Occupation. Liberia, which is nominally at peace following an end to its long civil war, is constrained by the distrust which was sown among the various ethnicities. From many countries in turmoil, such as Sudan, South Sudan, Yemen and Afghanistan, there has been a massive brain drain. It is possible, though far from certain, that if/when peace is restored, these emigrants will return, bringing their new experience and wealth with them to empower their mother country's development.

At a more transient level of political disorder which has affected food security, let's consider The Gambia. This small West African country has had a vibrant tourist industry, even under the repressive political regime of Yahya Jammeh. The presidential election of December 1, 2016 signaled a change of leadership, but the incumbent refused to accept the electoral result. The constitutional impasse, involving mediation by other countries, resulted in Jammeh eventually leaving the country on January 21,

2017. A military intervention, Operation Restore Democracy, under the Economic Community of West African States (ECOWAS) commenced the following day to police the transition of power. Adama Barrow was formally inaugurated President on February 18.

The mayhem caused by Jammeh's declaration of a 90-day state of emergency a day before his official mandate ended caused many tourists and diplomatic mission staff to leave, and many Western country governments issued travel warnings to their citizens against visiting The Gambia. Organized tours and individual bookings were canceled and incoming flights were discontinued. Hotel occupancy crashed to near zero, resulting in lower income of hotel and resort staff, due to loss of gratuities and reduced hours or layoffs. The allied hospitality supply and services industry also sustained greatly reduced incomes, and occasioned layoffs. The high season for tourism in The Gambia is February and March, so the hit was the more significant. Only since Barrow was installed have tourist numbers and revenues been recovering, following lobbying of European tour companies to persuade them that the country is once again safe for tourists. However, the holy month of Ramadan in 2017 started at the end of May, putting tourism recovery largely on hold as the "entertainment" scene is then minimal, as this keeps away many tourists, and after Eid Al Fitr that year it was too hot for many tourists, who would rather wait for the relatively cool season, starting in November.

So, for a whole year, a transient constitutional episode undermined the livelihoods and "accessibility" component of food security for those who depend on one of the main industries in the country. A major key to food security resilience is peace and personal security, and as indicated in Chapter 1, these parameters are mediated by humankind. There may yet need to be a peace and reconciliation initiative in The Gambia as there was under Mandela in South Africa, for the grievances of the Jammeh regime to be brought out into the open, for peace really to take root. People whose family members were killed are angry; the nation is angry.

Turning again to the global scene, angry voices have been plentiful in the realm of poetry, long before the advent of modern "social media" and their presence there. Unfortunately, those in the public sector at whom those voices have been aimed have often chosen not to listen. This must change if the diffuse World War III and entropy which we currently seem to be experiencing are to get no worse, let alone improve. Otherwise, what is the prospect for global food security and resilience?

A prime example of an angry voice from an "anthropological poet" is expressed nowhere better than in the poem "Identity card", by Palestinian Mahmoud Darwish. I was once nominated as "Professor of Poetry"—for a week—in a conflict resolution course for Israelis and Palestinians at the Salesian Pontifical University in Rome. Before the course, I asked the Palestinian delegates which poem from their "side" they would wish me to table for discussion. All agreed it had to be "Identity card", published in 1964, which caused outrage at the time among Israeli conservative hardliners. Its last verse in English translation is reproduced in Box 6.3.

In the last half of the book by Rose (1990), he considers "poetic anthropology", the voice of those who have traditionally been on the receiving end of formal ethnographic studies. Their anger spills over at being misunderstood, or worse, being dismissed as unimportant to engage with. One of the quotes in Rose's book is a poem called "I am the reasonable one" by the late Rosario Morales (1985). Morales identified herself as a New York Puerto Rican. In her poem she composes 11 verses on how reasonable she is to members of the white culture, which she regards as monolithic.

BOX 6.3 "Identity Card" by Mahmoud Darwish

... Write down
I am an Arab.
You usurped my grandfather's vineyards
and the plot of land I used to plow
I and all my children
and you left us
and all my grandchildren
nothing but these rocks ... so
your government
will it take them too as rumor has it?
So be it.
Write down at the top of the first page:
I do not hate people.
I steal from no-one.
However
if I am hungry
I will eat the flesh of my usurper.
Beware of my hunger
and of my anger.

Reproduced with permission from Wedde and Tuqan (1973) (Translated)

Then for the rest of the prose poem her reasonableness turns abruptly to anger—she "snaps", when structural integrity collapses under the increasing pressure applied to it—similar to what happens to a brick in a structural test in a building engineering laboratory (Box 6.4).

BOX 6.4 "I Am the Reasonable One" by Rosario Morales

... And I am angry, I will shout at you if you ask your venomous questions now, I will call you racist pig, I will refuse your friendship.

I will be loud and vulgar and angry and me. So change your ways or shut your racist mouths. Use your liberal rationality to unlearn your contempt for me and my people or shut your racist mouths.

I am not going to eat myself up anymore. I am not going to eat myself up anymore. I am not going to eat myself up anymore. I am going to eat you.

These poets talk into an anthropological, sociological and political space, a no-man's-land, a gulf between cultures that cries to be closed, yet sometimes is seemingly ever-widening (be it in Israel/Palestine, Myanmar, Syria, the United States or elsewhere). Today, as I write (September 17, 2017), I read of an interview with the BBC in which the current UN Chief is reported as saying that Aung San Suu Kyi has one last chance to stop the army offensive ... were I Burmese, I would have been incandescent at the ignorance displayed by that comment.[29,30] Such comments exacerbate rather than resolve the divide. The two institutional cultures, in this case in New York and Yangon, do "not share the same historical moment", as Rose says in a general context. The space between such cultures can be romanticized, but it can also be, and often is, angry (see Section 4.2.2.4). By contrast, the subtle persuasive approach undertaken by His Holiness the Pope before and during his visit to Myanmar in November 2017 was far more productive in influencing the country's authorities.

Poets have shown themselves good at expressing the feelings of the dispossessed and disenfranchised, perhaps less good at applying their art and sensitivities to proposing solutions and steering them through. This latter

---

[29]http://www.telegraph.co.uk/news/2017/09/17/aung-san-suu-kyi-has-last-chance-stop-rohingya-massacre-pulls.

[30]How different from the previous UN Secretary General's forthright and balanced statement to the UN Security Council on the situation in the Middle East of January 26, 2016 https://www.un.org/sg/en/content/sg/statement/2016-01-26/secretary-generals-remarks-security-council-situation-middle-east. Power in the country lies with the military not Aung San Suu Kyi.

role seems to me to be an area in which they could excel, working hand in hand with social scientists across the board toward social and political reform, applying a soothing balm to help heal deep-seated wounds, thereby facilitating an enabling environment in which negotiation could flourish.

As indicated in Section 4.2.2, there are so many cultural hang-ups which can get in the way of bipartisan discussion and understanding. For instance, in a conversation with an outsider, a Burmese may say "Yes" in response to a comment from the outsider, yet that likely does not mean "I agree with you", only "I have heard you". This is a very basic lesson for an outsider to learn when living in Myanmar. Within the "outsider" contingent, an Irish person may be the first to grasp this, for there is a parallel with the usage of "Yes" in conversations in the Republic of Ireland.

6.4.2 Absence of War

On the positive side, exemplifying the peace dividend, Rwanda continues to surge forward under the wise leadership of President Paul Kagame. The ethnic violence of previous decades has been consigned to history, following government peace and reconciliation initiatives between Tutsi and Hutu (under the National Unity and Reconciliation Commission), just as happened in post-apartheid South Africa under Nelson Mandela. The insurgency in northern Uganda has been routed, and that in northern Nigeria is being contained. In his inauguration speech of February 18, 2017 in Independence Stadium Banjul, President Barrow of The Gambia spoke strongly of the "victory for democracy" and empowerment of his people that could now come about, based on tolerance as the bridge between diverse tribes, all working together for the common good.[31]

In 1986 in Uganda, a series of brutal dictators was brought to an end through the people's liberation war under the National Resistance Army (NRA) which swept across the country from the west and drove the last of those dictators from Kampala, and out of the country to the east. The country was on its knees—I recall the piles of skulls that were collected from the bush in the "Luwero triangle" and heaped at the Bombo Road roundabout in Kampala, and the wooden splinters that served as staples for paperwork in local government offices. Museveni's Presidency since 1986 has encouraged all ethnicities to participate in the free economy, and all tribal groups have benefitted, and indeed the old divides between them are

[31]http://www.bbc.co.uk/news/world-africa-39011393 (accessed August 1, 2017).

all but gone. Interethnic strife, rife from independence in 1962 until 1986, is a thing of the past. The erudite and respected Prime Minister of the region of Buganda, Peter Mayiga, has told his fellow Baganda that they must become less proud, and stop thinking of themselves as better than other ethnicities in the country.

The improved economy may be one reason why peace has become established in Uganda.[32] After Museveni took power, personal security improved enormously. Following the NRA victory in 1986, soldiers lived in grass huts on the roadsides, to show people that the torture and detention centers masquerading as military barracks of former regimes were no more. People could stop living in fear, and need no longer sleep overnight in their banana plantations, afraid to sleep in their homes in case they were arrested or killed during the night. As a result of this transformation, businesses invested and people expanded their livelihoods and developed themselves. The economy flourished. The subsequent 10 years were a truly wonderful time to be living in Uganda. Democracy has been eroded somewhat since then, yet peace has been entrenched by so many people having a real stake in the buoyant economy. In other countries in Africa, democracy may be stronger, yet in some, peace is threatened by more effort being made to maintain or overturn the regime in charge than to boost the economy and improve public services.

The positive correlation between peace and the improved economy, food and other securities is beyond doubt. In the case of Uganda just described, it was peace and stability which brought about development, the "peace dividend" which empowered the people to develop themselves. Yet there are countries where the converse may be true, that peace will come about if the economy can be developed even under conflict, especially if the latter is localized or incipient rather than overt (Box 6.5). Facilitating both routes to development is surely a worthy goal.

Various foreign governments and donors have engaged with peace initiatives in many countries at war, some linking the peace process to providing development assistance. Sometimes peace is brought about by military intervention, as in Uganda and Rwanda already mentioned. In Mali (2012–14), peace was restored through French military intervention to quell an Islamist uprising against the Malian government, a move supported by the vast majority of Malians in its capital Bamako (though not of course by the Tuareg and Arab population of their Azawad ancestral homeland, whose

[32]In Rwanda similarly.

grievances about their marginalization attracted Al Qaeda support, and who all fled the towns into the desert or to neighboring countries, pursued by Malian government forces).

BOX 6.5 The Middle East Regional Agricultural Program (MERAP)

This development program evolved from a trilateral cooperative training program between Egypt, Israel and Denmark. The resulting agricultural cooperation program also involved Jordan and the Palestinian Authority (PA). Denmark took the intellectual and funding lead, drawing in collaborative funding from its partners, the whole being covered by one intergovernment agreement.

MERAP was implemented in three phases, from 1999 to 2016. Phase I was 1999–2005, Phase II 2005–10 and a delayed Phase III from January 2013 until the end of 2015. It proved impossible for Egypt to participate in Phase III, downstream of the "Arab Spring", so MERAP proceeded with the three remaining Middle Eastern countries. That this was possible is a remarkable tribute to the resolve of Jordan, Israel and the Palestinian Authority, and to program managers and participants on the ground. It is astonishing that it could have proceeded and succeeded against the background of the Middle East conflict, in particular the second *Intifada* (2000–05) and the "war" in Gaza in 2008–09.

The overall objective of Phases II and III was to contribute to the regional peace-building process. The subservient immediate objectives were twofold: (a) to increase and improve regional agricultural cooperation among partner counties; and (b) to increase farm income in a sustainable and environmentally-sound manner through enhanced effectiveness of agricultural extension services, transfer of know-how and strengthened farmers' organizations. That incremental income was generated as early as Phase II was in part attributable to the reciprocal trust among players built during MERAP I, and the capacity built during MERAP II.[33]

In other cases, negotiated peace comes through dedicated individuals at high political level, and covert facilitators, compelling combatant leaders to modify their views to give peace a chance. Northern Ireland is a case in point for the latter, though it is not a "developing country". Liberal-minded peace activists from Israel and Palestine who are addressing the long-standing Palestine–Israel dispute visited both the Republic and Northern Ireland in 2016[34] to see what can be learned from its successful peace process which could be applied to their own situation. For outsiders to offer themselves as peace brokers is good in principle. Resulting peace does not always hold

---

[33]The author was Final Evaluator for MERAP II and III.
[34]http://www.pij.org/details.php?id=1726 (accessed February 11, 2018).

though, the agreement on South Sudan in 2005 being a case in point—a complex situation with around 60–80 different ethnic groups, depending on classification, though the main protagonists are the rival traditional pastoral groups, the Nuer and Dinka. Yet it is surely worthwhile to try, and to keep trying, if conciliatory representatives of the groups in dispute invite or agree to such intervention.

Having myself had a minor covert facilitator role leading to the 2005 South Sudan peace agreement, and having worked with both Palestinian and Israeli peace activists in the Middle East, I opine that an outsider can indeed have a role to play which can be useful, if offered and delivered sensitively. I have observed during my 5 years' living in Palestine and Israel that it can be very difficult, if not traumatic, for Palestinians and Jordanians to sit with Israelis at the same table, there being so much pain just below the surface, from everyone's perspective.

Interventions involving outsiders can take many forms. One of three "peace poems" I had published in the liberal Jerusalem-based *Palestine-Israel Journal* (*PIJ*) (Ashley, 2014), which I called "Collateral damage", tried to capture how the current impasse is leading to the erosion of humanity on both sides of the Separation Wall, of both Palestinian and Israeli communities (necessitating massive humanitarian aid for Palestinians, and soup kitchens for impoverished Israelis in Jerusalem) (Box 6.6).

The *PIJ* is read by intellectuals and the political class on both sides of the divide. This poem was a small contribution from one person, inviting hardliners to approach more closely a median liberal view. With the concurrent international supportive or punitive interventions intended to foster a consensus, the promise of a better life for both will hopefully eventually hold sway to avert the very real risk of a regional war one day over the unresolved dispute.

My three peace poems in the *PIJ* were appreciated by a senior liberal in Israel's Department of Education, who tried to have two of them adopted as set texts in the school syllabus, in the year when the concept of "The Other" was a national theme. However, her hard-line boss ruled that these particular poems were a step too far, and would undermine the resolve of the youth in the Israeli military; they were thus not incorporated into the national school syllabus.

In 2004, over dinner with President Arafat in his residence in Ramallah, the Mukata, I mentioned to the President that I understood that if the Democrats won the forthcoming election in the United States, former President Bill Clinton would be the US government's emissary to the

BOX 6.6 "Collateral Damage" by John Ashley

It is a small bus, green and cream, number eighteen.
I know the drivers, been driven safely for years,
seen the fare rise time and again, now seven
shekels thirty, Ramallah to Damascus Gate.

After the Qualandiya sheep crush we
rejoin the bus and proceed. Yet today
we are flagged down for further scrutiny.
A flaxen-haired conscript mounts the rear steps,
her khaki authority greeted with
silent contempt. Boy soldiers strut outside
to impress her. Asked to do more than school
leavers should be, a crash course in bonding,
one Semitic race against another.
Collateral damage
to her self that can never be repaired.

I'm wearing a new linen shirt, present
from New York. As she moves up the aisle,
struggling to lift her gun, its range finder
catches my shoulder from behind, scags the fine
fabric. She did not intend to corrupt it,
knows not that she has, nor do I.
At home in the Old City though I find
flaxen threads distressed and distraught, broken
victim of circumstance, mute, unspoken.
Collateral damage
to my shirt that can never be repaired.

Middle East. Arafat replied with enthusiasm, "Why not?". He felt sure that Bill Clinton could deliver in an even-handed way, to both sides of the divide. Arafat was prepared to meet Ariel Sharon, Israeli Prime Minister at the time, but the latter would not see him. So a truly independent arbiter was welcome, from Arafat's point of view. The Democrats did not win that election, Bill Clinton could not work his magic and the impasse rumbles on, getting more intractable by the day.

It is necessary for such intercession parties to demonstrate their goodwill, independence and lack of bias. This requires potential arbiters to understand the views of all parties involved (without the need to agree with any of them). This in turn means that they need to grapple with the thinking and feelings of the contesting parties, to be prepared to start afresh by discarding the mindset of their homeland in Europe (say), and start understanding the perspectives of (for example) both the Buddhist and Muslim residents of Myanmar's Rakhine State, Palestinians from West Bank or Gaza and Israeli Jews, Nuer and Dinka in South Sudan and so on. This needs to occur despite the fact that no two people from any of these ethnic groupings will have an identical set of perspectives. Yet there will be some core commonalities, which need to be grasped if common ground is to be found between one group and the other as a basis for conflict resolution, through building trust and understanding, and greater tolerance of "The Other". This is a learning process from each side of the conflict, and not least learning by those outsiders who wish to assist.

Social and cultural anthropologists can be useful as facilitators in such negotiation, and to inculcate a wider understanding by some in the donor community, as can non-nationals who have lived in the given country for a long time. Another group of facilitators comprise those with well-honed diplomatic skills, like former US Presidents Clinton and Jimmy Carter and Senator George Mitchell, who are sufficiently wise to understand and weigh the disparate perspectives around the table.[35] It is not always possible to find a solution without a defeat on the battlefield, but it is worth trying.

Within the country's nationals, as in the case of Northern Ireland, one ethnic or religious group needs to learn how to accept another point of view, to learn to acknowledge, even accept, "The Other"—the other person, the other point of view. Quite remarkable transformations can result, which once seemed impossible. An example of the deep friendship which can develop between formerly vehement protagonists on different sides is that between the late Protestant firebrand the Revd. Ian Paisley and Martin McGuiness of the Irish Republican Army (IRA), both latterly referred to as the "chuckle brothers". Another unlikely friendship pairing that came about was that of President F.W. de Klerk and Nelson Mandela in South Africa, which resulted in the end of the apartheid era of government. These friendships came about because the men decided that discussion and sharing ideas on the way forward was the route to take to combat fear, to roadmap a future out of a pointless stalemate, and rapprochement and friendship was an outcome that empowered and facilitated negotiation.

---

[35]Each has written passionately on the "peace process".

6.5 OTHER PRIORITIES TO IMPROVE RESILIENCE

Related to the preceding "peace" priority is the priority of natural disaster risk reduction (DRR). One specific example is given to illustrate why DRR is a priority for many coastal communities in the developing world, in this case in Myanmar.[36,37] The number of people living in areas threatened by storm surges in Myanmar is likely to increase from 2.8 million to 4.6 million by 2050, according to the multidonor Livelihoods and Food Security Trust Fund (LIFT) program. The strengthening and building of embankments to protect rice paddies and coastal villages, though regarded as a huge success under the Tat Lan project, can offer only a short-term provisional answer to climate change and natural disasters. A long-term solution needs to be found for those living in coastal areas subject to cyclones and rising sea levels, thereby building resilience for the community at large, against poverty, hunger and conflict. A project proposal formulated by the Geneva-based non-profit association Displacement Solutions "Developing institutions to prevent climate displacement and land conflict: The Myanmar National Climate Land Bank" (October 2016) has such a long-term view. The proposal is that the feasibility of set-aside land banks be considered, to provide living and livelihood space for the millions who will inevitably be displaced by extreme weather events and climate change over the forthcoming decades. This will help forestall future conflict resulting from climate internally-displaced persons (IDPs) being seen and treated as a threat to the host communities to which they flee.

Other equally-deserving candidates for attention as food security priority challenges have been addressed in the author's previous book, such as cultural and social norms and practices as they influence nutritional sufficiency (Section 6.2). Social safety nets have also been addressed throughout the earlier book, especially in Section 5.4.2 and Case Study 1 (The Productive Safety Net Program in Ethiopia—the PSNP) on the companion website. In the current volume, the PSNP has been referred to, and a consideration given to Qali Warma in Peru in Section 6.2.5.2. Suffice it to expand here on what was said in Section 5.3.2 of the earlier book, namely that the main

---

[36]Section 6.9 in the earlier book, and Figure 6.2 therein, show the coastal cities in Africa at particular risk, such as Port Sudan.

[37]As I write (September 9, 2017), much of Florida is under water from the storms and tidal surge associated with Hurricane Irma, and the humanitarian crisis facing the Bengali Muslim population in Rakhine State of Myanmar is extreme; hundreds of thousands of Muslim residents have fled to Bangladesh over recent weeks.

goals for social safety nets (promoting social inclusion, better livelihoods and resilience to shocks) have already been achieved in some parts of the Middle East and North Africa (MENA) region, despite many challenges (Silva et al., 2013). Before the Arab Spring, many governments in MENA tended to rely on a distribution system that protected the population against destitution through universal subsidies of basic consumption items, which guaranteed affordable access to food and fuel for all its citizens, irrespective of their needs. This system's response to crises was to scale up these subsidies or to provide more and better-paid public employment for middle and upper classes. While understandably popular, neither is sustainable, nor do they empower citizens to engage with better livelihoods.

A policy priority for many current oil- and gas-rich nations is food self-sufficiency and better resilience through diversification, against a time when their non-renewable sources of wealth have been depleted. For example, Brunei Darussalam imports more than 80% by value of its food requirements, 75% of that coming from other Association of Southeast Asian Nations (ASEAN) countries.[38,39] Farming has become a part-time business for most rural families, owing to the availability of more lucrative forms of employment. Since 2008, with the food crisis conditioning political and economic agendas across the world, trading of food and food products has become the major concern for Brunei. Through the ASEAN regional grouping, particularly the ASEAN Food Security Reserve Board and the Agricultural and Food Marketing Association for Asia and the Pacific (AFMA), Brunei is working to modernize the rice trading mechanism in the Asia Pacific region to better stabilize the international market price. Simultaneously, the government, farmers and private sector are endeavoring to improve resilience, with the Department of Agriculture and Agrifood taking the lead.

By 2011, national production of rice, Brunei's main staple, totaled just 1,480 tons, accounting for only 4.44% of the 33,315-ton annual demand then; self-sufficiency in fruits and vegetables stood at 15.4% and 62.8%, respectively, with less than 1% for beef.[40] Brunei is making strong efforts to improve its self-sufficiency in local rice, though it has only a potential 5,000 ha of suitable land available. Since 2011, productivity of rice has

---

[38]https://gain.fas.usda.gov/Recent%20GAIN%20Publications/Exporter%20Guide%20
Annual%202016_Kuala%20Lumpur_Brunei_11-16-2016.pdf (accessed September 22, 2017).
[39]https://www.wto.org/english/tratop_e/tpr_e/s196-00_e.doc (accessed September 22, 2017).
[40]http://www.oxfordbusinessgroup.com/brunei-darussalam-2013 (accessed December 5, 2015).

Photo 6.3 Motorbike taxi rider and mechanic, Banjul, The Gambia, both important sources of income in developing countries, especially for the male youth (June 1, 2017).

Photo 6.4 Mrs. Ndye Corr, in Banjul, The Gambia, with samples of local fruit juices and dried flowers (sepals) of the Hibiscus flower ("sorrel" or "karkadé"). Having qualified in fruit processing, she runs the business from home, employing 10 people. Women farmers bring their produce to her, which she then processes, packages smartly and markets (June 3, 2017).

increased considerably due to better varieties and management, but there is need to attract more foreign firms and foreign direct investment to improve national productivity and production further. In land-based agricultural activities in general, low capacities and outdated practices have blunted the effectiveness of government reform initiatives. For instance, most of the rice is currently rainfed rather than irrigated, and the agricultural labor force is largely externally-sourced. Meeting such challenges in a sustainable way will continue to define the agriculture and agribusiness sector for now.

Economic access to food is the main direct cause of food insecurity in developing countries. Anything that can be done to promote agribusiness (crops and livestock), SMEs, sustainable natural resource harvesting, and indeed any kind of legal income-generating activities, including policies which better enable the local trading environment, constitutes a priority. Photos 6.3–6.8 show typical examples of these activities.

Photo 6.5 Value-adding attractive packaging and display of various locally-grown fruits at a roadside stall, near Al-Ramtha, NW Jordan (July 12, 2016).

Photo 6.6 Yarn spinner at Tibetan Refugee Self-Help Centre in Darjeeling, West Bengal, India (April 2013) (see http://www.tibetancentredarjeeling.com). *(Courtesy of Martin Tayler.)*

Photo 6.7 Cobbler at Ghoom, near Darjeeling, India (April 2013). *(Courtesy of Martin Tayler.)*

Photo 6.8 Wood carvers at the Zorig Chusum Arts and Crafts Center, Thimphu, Bhutan. Boys and girls who are deemed to be not academic are eligible for a 6-year training in various arts and crafts, such as textiles, painting, sculpture and wood carving) (April 2013) (see http://www.tourism.gov.bt/about-bhutan/Arts-crafts). *(Courtesy of Martin Tayler).*

REFERENCES

Allen, T., 2015. Moving Beyond Agriculture: It's Food that Matters! OECD Insights: Debate the Issues. Sahel and West Africa Club/OECD Secretariat, December 4.

Arnould, E.J., 1989. Anthropology and West African development: a political economic critique and auto-critique. Hum. Organ. 48 (2), 135–148.

Ashley, J., 1999. Food Crops and Drought. The Tropical Agriculturalist Series, Macmillans, UK, pp. 89–90.

Ashley, J., 2014. Collateral damage. Palestine-Israel J. 19 (3), 119.

Bate, S.P., 1997. Whatever happened to organizational anthropology? A review of the field of organizational ethnography and anthropological studies. Hum. Relat. 50 (9), 1147–1175.

Berg, A., 1987. Malnutrition: What can be Done? Lessons from World Bank Experience. World Bank, Washington, DC.

Bliss, F., Neumann, S., 2008. Participation in International Development Discourse and Practice. "State of the Art" and Challenges. University of Duisburg-Essen, Institute for Development and Peace (INEF), Germany. Report 94/2008. 72 pp.

Coleman-Jensen, A., Rabbitt, M.P., Gregory, C.A., Singh, A., 2017. Household Food Security in the United States in 2016, ERR-237. U.S. Department of Agriculture, Economic Research Service Report 237. 44 pp. https://www.ers.usda.gov/webdocs/publications/84973/err-237.pdf. Accessed October 19, 2017.

Collinson, P., Macbeth, H. (Eds.), 2014. Food in Zones of Conflict: Cross Disciplinary Perspectives. Berghahn Books, New York. 235 pp.

Ferguson, J., 2006. Global Shadows: Africa in the Neoliberal World Order. Duke University Press, Durham, NC. 258 pp. (p. 87).

Gellner, D., Hirsch, E. (Eds.), 2001. Inside Organizations: Anthropologists at Work. Berg, Oxford.

George, S., 1976. How the Other Half Dies: The Real Reasons for World Hunger. Penguin, London. 352 pp.

George, S., 2004. Another World is Possible If. Verso, London. 268 pp.

Heidhues, F., et al., 2004. In: Development strategies and food and nutrition security in Africa: an assessment. IFPRI, Washington DC, 2020 Disc. Paper 38, December 2004. 60 pp.

Juma, C., 2011. The New Harvest: Agricultural Innovation in Africa. Oxford University Press, New York. 296 pp.

Kirschenmann, F.L., 2013. Anticipating a new agricultural research agenda for the twenty-first century. In: Albala, K. (Ed.), Routledge International Handbook of Food Studies (2012). Taylor & Francis, Abingdon, pp. 364–370 (Chapter 33).

Messer, E., Cohen, M.J., D'Costa, J., 1998. In: Food from peace: breaking the links between conflict and hunger. Food, Agriculture and the Environment Discussion Paper 24, IFPRI.

Moock, J.L., 2011. In: Network innovations: building the next generations of agricultural scientists in Africa. ASTI/IFPRI-FARA, Conference Working Paper Under Human Resource Development for Agricultural Research & Development, Accra, Ghana, December 5–7, 30 pp.

Morales, R., 1985. I am the reasonable one. In: Iain Prattis, J. (Ed.), Reflections: The Anthropological Muse. American Anthropological Association, Washington, DC. 283 pp. (pp. 201–202).

Nanda, S., Warms, R.L., 1998. Cultural Anthropology, sixth ed. West/Wadsworth, USA, pp. 350–368.

Ó Gráda, C., 2015. Eating People is Wrong, and Other Essays on Famine, Its Past and Its Future. Princeton University Press, New Jersey.

OECD, 2016. Territorial Reviews: Peru 2016. p. 203.

Ofir, Z., Swanepoel, F., Stroebel, A., 2014. On the road to impact and resilience: transformative change in and through AET in sub-Saharan Africa. In: Swanepoel, F., Ofir, Z., Stroebel, A. (Eds.), Towards Impact and Resilience: Transformative Change in and Through Agricultural Education and Training in Sub-Saharan Africa. Cambridge Scholars, Newcastle, UK, pp. 462–489 (Chapter 17).

Pottier, J., 1999a. Anthropology of Food: The Social Dynamics of Food Security. Polity Press, Cambridge. 230 pp.

Pottier, J., 1999b. Food security in policy and practice. In: Pottier, J. (Ed.), Anthropology of Food: The Social Dynamics of Food Security. Polity Press, Cambridge. 230 pp. (Chapter 2).

Rose, D., 1990. Living the Ethnographic Life. Qualitative Research Methods Series 23. Sage, Newbury Park, CA. 64 pp. p. 12.

Sen, A.K., 1981. Poverty and Famines: An Essay on Entitlement and Deprivation. Clarendon Press, Oxford. 257 pp.

Shore, C., 2012. Anthropology and public policy. In: Fardon, R., et al. (Eds.), The SAGE Handbook of Social Anthropology. vol. 1. Association of Social Anthropologists of the UK & Commonwealth, London, Los Angeles, pp. 89–104 (Chapter 1.6).

Shore, C., Wright, S., 1997. Anthropology of Policy; Critical Perspectives on Governance and Power. Routledge, London.

Silva, J., Levin, V., Morgandi, M., 2013. Inclusion and Resilience: The Way Forward for Social Safety Nets in the Middle East and North Africa. World Bank, Washington. 272 pp.

Sridhar, D. (Ed.), 2008a. Anthropologists Inside Organisations: South Asian Case Studies. SAGE Publications, New Delhi.

Sridhar, D., 2008b. The Battle Against Hunger: Choice, Circumstance, and the World Bank. OUP, New York. 229 pp.

UNEP, 2007. Agriculture and the environment. In: UNEP, Sudan: Post-Conflict Environmental Assessment. United Nations Environmental Program, Nairobi, pp. 158–191 (Chapter 8).

Waldman, A., 2002. India's poor starve while wheat rots. New York Times (December 2).

Wedde, I., Tuqan, F. (translators), 1973. Selected Poems of Mahmoud Darwish. Carcanet Press Ltd., Manchester, UK, p. 25.

Weis, T., 2013. The Ecological Hoofprint: The Global Burden of Industrial Livestock. Zed Books, London/New York. 188 pp.

World Bank, 2007. Cultivating Knowledge and Skills to Grow African Agriculture: A Synthesis of an Institutional, Regional and International Review. World Bank, Washington, DC. 117 pp.

World Economic Forum, 2013. Achieving the New Vision for Agriculture: New Models for Action. Report Prepared in Collaboration With McKinsey & Company, Davos, Switzerland. 32 pp.

Wright, S., 1994. Anthropology of Organizations. Routledge, London. 217 pp.

CHAPTER 7

Building the Change Management Team and Approach

Contents

7.1	Advantages of Working Well as a Team	112
7.2	Getting It Right as a Team	115
7.3	Getting It Wrong as a Team	116
	7.3.1 Territorial Integrity	116
	7.3.2 The Start of Boko Haram	116
7.4	Institutional Perspective on Change Management	117
7.5	Program Implementation	119
	Reference	123

It is a truism that one person can make a difference on his or her own, yet far more can be done by a good team, from the planning stage through implementation and beyond. Ask any Premier League soccer team manager. Synergies can be exploited and a critical mass of effort created to have a measurable impact on the resilience constraint being addressed. This is well-attested by the global attainment of Millennium Development Goal 1, Target 1c, as pointed out on the previous book's companion website.

Yet a team can also be dysfunctional, if each individual within a group competes to be the alpha male or female, making it difficult for the designated team leader. In countries that are socially fractured, for example, Palestine, Liberia and some countries in Central Asia, achieving good team cohesion can be a particular challenge. And yet, as British Prime Minister Churchill once said (paraphrasing playwright Oscar Wilde): "There is only one thing worse than having allies, and that's not having them". Working to build the team can involve much patience and compromise. It can be particularly difficult when team members have not met before, and the period over which they interact is short.

Individuals are needed with knowledge and skills, and not least, wisdom. Those who work in the field with the "beneficiary group" must have the respect of the local community, and authority because of it. I have to

Human Resilience Against Food Insecurity
https://doi.org/10.1016/B978-0-12-811052-2.00007-X
111

say too that the best candidates for "technical" positions in the field are not necessarily those who have the best qualifications or the longest list of publications. Such people often have little experience in the "field", are not used to problem-solving within marginalized communities, and do not have a feel for the private sector entrepreneurial spirit, where there is no salary, as income is rooted in results. I read a job specification from an NGO recently which included the text: "What you are qualified in doesn't matter. What you are passionate about does". The best applicant for a job advertised in a project with which I was working in the 1990s was someone without a degree at all. He shone from the moment he was asked to say something about himself. His confidence, his vision and dependability came across strongly. He was the unanimous choice of the selection panel.

In the case of leaders identified from the beneficiary community, such respect and authority needs to be anchored in their previous achievements in that community. The need to work hard in a dedicated way is an important aspect of attaining that respect. As discussed in the previous book, women have drawn the short straw on this, usually deserving, yet not being accorded, respect from their menfolk. The presentation above can be developed further by being partitioned into several headings, such as those below, all with a focus on building resilience against food and nutrition insecurity.

7.1 ADVANTAGES OF WORKING WELL AS A TEAM

While formulating potential investment projects in Liberia in 2011, I had a Ghanaian-American colleague who number-crunched their gross margin analyses. His training had included an MBA from Harvard, and he related to me something about the course that particularly interested me. At the very start, the students were set tasks. These were so daunting that, without prompting from the course tutors, the students decided among themselves that they needed to cooperate if they were to complete those tasks. This was what the course tutors had wanted to happen, and comprised Lesson 1, namely "to achieve results, cooperate!".

On offer is support from one member to another, with each member contributing something to the team's output, that output being more than the sum of the component contributions. Often this relates to a multidisciplinary team cooperating, each bringing his or her own specialization and comparative advantage to the table and the field. By training, my Ashanti colleague is not an agricultural production/productivity specialist and I am not an economist. Yet together we achieved in Liberia, through the synergy generated.

In another situation, in Eastern Africa, I recall being grateful to the economist on the team in the Uganda Wheat and Barley Program, for although my counterpart barley agronomist and I proved that applying foliar fungicide spray to a barley crop on Mount Elgon significantly increased grain yield, our economist colleague showed that the benefit was not one which was profitable, so should not be part of our recommended barley production package for that site.

Sometimes, *ex-officio* team members materialize serendipitously. In 2017, I was conducting an evaluation of a food security program in The Gambia. On a related up-country field trip I visited a participating school to review its school feeding program. The school had a small garden on its compound wherein it was trying to grow vegetables to help provide lunches for the pupils. Lamin Gitteh, the "garden master" on the school staff, took me to the garden, which abutted The Gambia River at Janjanbureh (Georgetown), 175 miles upstream of the river mouth at Banjul, and was subject to estuarine tidal rise and fall. He explained that there was never a night when hippopotami did not come out of the river and graze on/trample over the vegetables, and he could not think of a way to prevent it. I observed a solitary break in the mangrove bushes lining the bank, and suggested that a derelict canoe (*pirogue*) be dragged off the river bank, floated to the gap in the trees and then scuppered in the shallows to close the gap. That should serve as an adequate bar to entry of hippos into the garden. The headmaster and garden master were amazed that such a simple solution could resolve their dilemma, one which neither had thought of himself. I muse that this was the first time I have been called upon over 40 years in the development field to devise a hippo deterrent strategy—lateral thinking at work (Chapter 9).

Another similar *ex-officio* intervention I made was in 1997 when staying at a low budget lakeside hotel at Pokhara, Nepal, when I was natural resources adviser to two districts in the west of the country. While sitting in the restaurant in the evenings I watched how the staff worked, and started to make a list of organizational and ergonomic changes that could be made to improve efficiency and cost-effectiveness. When the list was quite long, I invited the hotel manager to my table one evening and discussed the management points with him. He was so delighted, he told me that he would implement all the changes I'd suggested, and that I would not be billed for accommodation for the rest of my stay!

This story reminds me of an evening in Uganda, my team being taken out for dinner in Jinja by the late Assey Mukasa, Managing Director of Uganda Grain Milling Company, where I was then working. Assey surveyed the disposition and actions of the restaurant staff and mused that the work being done by the five staff would in Europe be done by one waiter or waitress, and that person would still have time to spare! Therein lies one reason why one country has developed while another is still developing. In a word: "management".

This point was made by President Museveni in his early days in that role, when visiting industries in Jinja, which is a town with the second largest economy in the country. He visited one factory on the shore of Lake Victoria, and asked its managing director: "What is your annual turnover?". The MD could not provide an answer, whereupon President Museveni turned to his secretary and said, "Secretary, make a note that the Managing Director of (abstracted) Uganda Ltd does not know his annual turnover". What a put-down! Hopefully, lessons were learned, by the MD concerned and by others through the telling of the tale at golf clubs across the country. Indirectly, the President had become part of "the team".

Another window on *ex-officio* members of the team comes to mind from my time in Liberia. My favorite restaurant in Monrovia is a hub for private entrepreneurs. I found it a wonderful place to patronize while I was developing agribusiness proposals for the Ministry of Economy and Planning, to test ideas informally against investors and business people. CMAM clinics are another venue at which I learn so much about the nutritional health of the community and visits to local vegetable and fruit markets are instructive, about the quality of produce available and its origin. Clinic staff and market vendors are thereby *ex-officio* members of the team with whom to network (Photos 7.1 and 7.2).

Photo 7.1 Market scene in Punakha, Bhutan (April 2013). *(Courtesy of Martin Tayler.)*

Photo 7.2 Market scene in Punakha, Bhutan (April 2013). *(Courtesy of Martin Tayler.)*

7.2 GETTING IT RIGHT AS A TEAM

In the interests of building a development effort to "fire on all cylinders" to bolster resilience in the community, a well-oiled machine is necessary, each part engineered to be fit-for-purpose to complement the role of other parts. Away from this engineering metaphor, how is it best to get one group of development actors to engage with other groups? This needs to be addressed through "inspection", "reflection" and "analysis", manifesting itself as "observe, listen, study, enjoin, be humble". Some lateral thinking does not go amiss either. An example in the author's experience is related below, which could also have been included in Chapter 9 on Lateral Thinking.

While formulating the Palestinian National Food Security Strategy in 2005, there was a need to be "participatory", something that was a fairly new modality for the public sector (see Chapter 1.8 in the earlier book). The stakeholder analysis showed which were the key stakeholder groups within the public and private sectors, civil society and donors. The challenge then was how to get the various stakeholder representatives to talk with and listen to one another, because normally they don't. Based on observation, participatory discussions in workshops were adjudged insufficient to ensure sustainability and joinedup'ness of the proposed Strategy.

I sought something that all would enjoy out-of-hours, at a workshop to be held at an hotel in Jericho in the Jordan Valley. I hit on the idea of an *ud* (lute) player, a musical instrument to which I was partial myself. So

I arranged to bring a lutist from Jerusalem, and he gave a highly participatory recital on the evening before the two-day workshop plenary session. To say he was a "success" is to underestimate his impact. Even before the first formal session, there was a "demand" that the *ud* player be brought back for another night. This was arranged, though it broke the workshop budget. The workshop was a huge success, people who had never met before interacting as long-lost friends. The scene was set for compromise and agreement. This was how the National Food Security Strategy was formulated and agreed, the first time a national strategy was devised in a participatory way since the Palestine National Authority was formed in 1994. The power of music ....

7.3 GETTING IT WRONG AS A TEAM

7.3.1 Territorial Integrity

When on trek in eastern Nepal, we pitched camp on the recommendation of the porters in a thicket next to the Arun River. I noted a strange odor but dismissed it as nothing to be concerned about. During dinner around the camp fire on the first evening, one of our party suddenly stood up and shouted, "Tiger!". We all stood up as one, faster than anyone of us had ever stood up before or since, to hear something bounding away through the bushes. The one who had raised the alarm had seen eyes twinkling in the firelight, just beyond the dining party circle. The odor that I had noted was the marking scent of a big cat—probably a leopard rather than a tiger. We had camped on its territory, though we were safe enough because of our numbers.

7.3.2 The Start of Boko Haram

An instance of project management getting it wrong was in northern Nigeria. While conducting a review of agricultural extension systems across the country's States at the turn of the millennium, the team did a lot of traveling. I was one of a subteam to travel from Enugu, Igboland to Borno State in the northeast of the country. To save money, the project manager hired for the purpose an old saloon car with driver. It broke down four times on the journey, each time needing running repairs by the side of the road. We should have reached Maiduguri before dusk, but because of the delays we were still far short of it by then. Just after dark we found ourselves ambushed by gunmen. Ours was not the only vehicle "inconvenienced"— our car boot was stripped of everything, including my laptop and a lot of project money I'd (unwisely) been asked to carry. My colleague was shot in the chest with an arrow, which fortunately struck a rib bone and did not penetrate the thoracic cavity. I was beaten on the back as I called for the driver to bring the key to open the boot.

The language the robbers used was not known to the Nigerians in the car, indicating they had crossed the northern border to conduct the raid. Earlier in the day we had given a lift to a sick soldier whom we encountered at the roadside, and dropped him at a village health center, just before we reached the hold-up roadblock. Had the armed soldier been in the vehicle when we were held up, I would not be relaying this tale now. So, a litany of mistakes—to hire a substandard vehicle, to move with a lot of cash, to pick up a soldier ... The raiding party disappeared into the sand dunes with their booty and we made our escape, stopping at the next police station to make a report. We heard later that the raiders were pursued and wiped out soon afterwards by a military unit dressed in mufti. This was likely an early episode of the Boko Haram insurgency.

7.4 INSTITUTIONAL PERSPECTIVE ON CHANGE MANAGEMENT

The management approach and system whereby an intervention is undertaken is crucial to its success. It is important that checks and balances are in place during implementation, through adequate monitoring, reporting, steering committee meetings (inclusive, with appropriate seniority of membership and follow-up), smooth funding flow, monitoring and evaluation. For sustainability purposes, a "project" budget line ideally needs to be incorporated into State finances; without this "ring-fencing", there may be no postproject continuity. For this to happen, from project identification stage there needs to be a **government** buy-in (one or more Ministry, Department or Agency), for which various levels of executive and administrative authority need to be kept involved and informed, in a scenario in which competition for funding is high. The need for government buy-in is exemplified in a report by the World Bank, noting specific ways in which the Nigerian government needs to act now in relation to climate-resilient development (Cervigni et al., 2013). Otherwise, there are worsening prospects for food security countrywide, especially in the north and southwest.

Other institutional stakeholder groups comprise **international organizations, development banks** and **donors**; **regional groupings** such as the South Asian Association for Regional Cooperation (SAARC), the Economic Community of West African States (ECOWAS) and the Southern African Development Community (SADC); international **NGOs** (INGOs), local NGOs (LNGOs) and Civil Society Organizations (**CSOs**); and **consultants, university** research and teaching institutions, and **vocational schools**. It would be nice to think that these players work together seamlessly and with synergy. On occasions they may do, but often they do not, for reasons of "protecting rather than diluting their patch of influence and funding sources", or simply due to not knowing or understanding what the other players do. Within-category there is a lot of competition too, for similar reasons.

Examples of this I've noticed, which are counter to the notion of building resilience in a country, include an academic Food Security grouping chaired by professors with very specific technical disciplines, yet with none of them having a general overview competence, grounded in years living "at the sharp end" in developing countries. Such a person on the overall management team (indeed more than one), it seems to me, is essential to construct the in-house overview, to help pinpoint the real needs "out there", to identify those which can be addressed empirically, to prioritize them in a demand-led not supply-led modality and to formulate the integrated roadmap.

Just as agronomists need economists (see Section 7.1), so plant breeders need anthropologists and biochemists. The anthropologist can guide the breeder in developing "varieties" that will be welcomed by farmers, and help to promote them in the community once they are available. The skill sets of the two disciplines are complementary. And in a biofortification initiative, biochemists can screen existing lines of a crop for high values in Provitamin A or lysine (say), or whatever trait is needed to target micronutrient deficiencies in the community, as advised by nutritionists; these lines may be suitable for release directly or be used in a breeding program. Agricultural research workers and extension workers have traditionally worked as if they were distinct subspecies of the human race, this to be regretted. It is imperative that they work together, and be farm-based with much research being undertaken on-farm under farmer management, for example, the PVS initiatives mentioned in Section 8.7.

Another window on the "silo mentality" is the reluctance of one international research body I know to produce learned reports with contributions from the groups which work on the ground with end beneficiaries. I have also come across project steering committees without representatives from the end beneficiaries, resulting in insufficient holding-to-account of implementing agencies. Furthermore, I have noted donor agencies which do not interact anywhere near sufficiently with local NGOs and national opinions in general, believing that the thinking patterns and skill sets they have brought from developed countries are sufficient.

There are also international NGO staff who do not appoint nationals as running mates, involving them in report/proposal writing and crucial decision making; such INGOs thereby do not fulfil their moral mandate, even if it is not written into their constitutions, that they strive to make their international staff redundant because they have trained nationals to do their work—technical, financial, organizational, operational and management. Such pairing of staff is clearly in the interest of sustainability and building resilience in the national community, and operating thus should, in the author's view, be mandatory if an INGO is to be selected to implement a project under a Call for Proposals (CfP) tender. Capacity building of local NGOs is a crucial tool for INGO staff to undertake, not least to help them construct proposals for funding. What is happening currently is that INGOs write the proposals for such CfP funds, or even subcontract the task to international consultants. The danger then is that the whole tenor of such proposals is oriented towards the needs of the INGOs and donors rather than the needs of the beneficiary communities.

Donors are in part to blame, laying down such exacting requirements for proposal content and implementation that a local NGO feels totally intimidated and its confidence undermined; to the contrary, writing and implementing a program should build that NGO's confidence. I recall reading the final report of an LNGO in Zambia saying that though it "enjoyed" implementing the given project, next time it implemented an EC-funded program it would be sure to recruit someone with a degree in EC procedures! I passed on this sentiment to EC HQ, Brussels. UN organizations can similarly be mired in protocol which acts as a dragline on field implementation; I recently came across one which demanded that an implementing NGO complete a detailed procurement specification for some donkeys it needed for a project involving animal traction.

7.5 PROGRAM IMPLEMENTATION

The implementation team, be it consultant- or NGO-managed, needs to satisfy various levels of client. It is not only the donor agency, but also the national government agency involved with its local line agency administrations, local leaders (civil, traditional and religious), and the HQs/local offices of the implementer's employers—consulting companies, INGOs, etc., and, most importantly, the end beneficiaries.

To reach the hearts of the latter group it is helpful to leverage on local dignitaries and social institutions, such as the imams and priests, local politicians and administrators, where this is deemed appropriate. The latter categories may not necessarily be appropriate if their personal interests outweigh their public responsibilities. Local radio stations and print media can prove their worth as a vehicle for raising awareness and releasing project-related messages.

Most of the prioritization should have taken place during the program formulation, yet it may need adjusting in due course. The mid-term review of the project offers a suitable opportunity for such modifications to be tabled and changes made. One issue that may come to light is whether the project is sufficiently resourced—very often the goals and achievements in the logframe are hopelessly unrealistic, with insufficient staff or time to achieve. This may be because the project document has been written by someone without knowledge of local conditions and related constraints, and all the reasons why something cannot be done on time—like the dearth of local consultants and/or that they are university lecture-oriented rather than practically oriented.

A common occurrence too is that Terms of Reference (ToRs) for a project formulation or evaluation have been written by someone who may well never have done, and could never do, the work himself or herself which is being advocated therein. In terms of time needed to do a job well, a rule of thumb in countries like Afghanistan, or in the Gaza Strip say, is "Think of a number, then double it". In other words, because of the multitude of reasons which may impede progress, twice the number of days may be needed as have been allocated in the ToRs. Consultants usually have to work seven days a week anyway, though they are probably only paid for five, and more time is needed to write the Final Report than the few days commonly allocated in the ToRs.

Hopefully, the problem tree will have been identified at the outset of program formulation, the apex constraints in the community laid bare and appropriate decisions made on which ones are suitable to address in the timescale for which funding is available. A Trust Fund arrangement provides excellent multiyear funding security, such as the PSNP in Ethiopia; short-term projects involving agriculture often do no more than start a process of change which is not sustainable, for it may take the first year just for cumbersome procurement arrangements to be worked though. Project gains may fall apart postproject if an exit strategy has not foreseen an aftercare service, to allow for on-the-job mentoring for those national staff trained during the program to build their confidence, and for provision of spares for equipment or vehicles to be purchased.

A flat management style is preferable to the traditional pyramidal style, with the project manager wearing his or her authority lightly, leading by example not edict. The Law of Subsidiarity should pertain, where authority is delegated as far down the hierarchical management chain as possible, this facilitating optimization of local solutions to local challenges. To actually enact this seems to be particularly difficult for some international organizations, which are built on the pyramid chain of command. This can lead to inordinate delays. A trivial example was when I was evaluating a project in West Africa, but could not get the country representative of a UN organization to agree to my riding to site as a passenger in his organization's vehicle. His HQ had to authorize it, he said, and the authorization was not forthcoming, despite my willingness to sign a waiver in the event of an accident. To save delay I found my own way to site by thumbing a lift on a hospital staff bus and finishing the journey on the back of a motorbike. I related the incident to the country representative of that same organization in another country and she was adamant that the decision to authorize could have been made locally. Another example was a project formulation in Central Asia, when the UN agency involved insisted it must first write a letter to government organizations requesting meetings. The agency was so mired in protocol that it did not produce the letter, so my national consultant colleague used his networks within government and we had all the meetings we needed, much to the surprise of the UN agency concerned.

Fine examples of such flat-style light-touch management I saw during 2017 included that by the WFP in The Gambia (see Section 6.2.5.1), and by the INGO HelpAge in Myanmar, where it is building the capacity of the Ministry of Social Welfare, Relief and Resettlement (MSWRR) in social protection issues. International managers with these two organizations have informed themselves of demand-led needs in the field, and the political and social environment which shapes realities there, working in tandem with local NGOs/CSOs who are close to the beneficiaries. The managers lead their national staff by example, encouraging them at every turn, putting them in the spotlight to make presentations, guiding not dominating while offering mentoring as necessary from the sidelines. This approach is an effective tool for sustainability of the initiatives being implemented, a perfect exit strategy conducive to eventual resilience within vulnerable communities.

At the start of my career as a consultant I asked the late Frank Sole, the MD of my company, if I might pursue a particular initiative. He could tell both that I really wanted to do so, and yet that I was concerned that he might say, "No". He said, "John, if you don't want the answer 'no', don't ask me the question". In other words: "Use your initiative—if it works you'll be praised, if it doesn't you'll have to answer for it". Wonderful advice. If you think you are onto a winner, follow your dream. Like a sportsperson, if your idea works and you "score", you become a hero, even a legend. If it doesn't, you'll be "hung out to dry". What a thrilling challenge.

Other good management tools for the project manager to follow, and share with his or her colleagues, include understanding that it is not a weakness to say, "I need help in completing this task", when one is overwhelmed with many tasks that one person cannot possibly have the time or knowledge to complete. This can often be preempted by the project manager asking his or her staff, "Have I given you enough for you to do your job?". I recall recommending to a staffer in a donor office in West Africa that he needed help in doing his job, as he was clearly overstretched and not having sufficient time to monitor the activities of NGOs in the field. He was such a pleasant person, and could serve as the "good cop", encouraging the NGOs, yet there was also a need for a "bad cop" who would police the activities to ensure that the performance was up to standard. The staffer objected to the notion that he needed help. Later, however, I was pleased to note that on the basis of my recommendation, extra consultant TA had been drafted in to assist staffers in that donor's country offices.

My younger son Rufus has recently started a career as a building surveyor for a large construction consulting company in London. In 2017 his company, along with a few others, signed a training partnership with the Royal Engineers (RE) Corps of the British Army. Both the RE and his company have a prime interest in physical infrastructure, and the company thought it worthwhile to seek the RE's help in team building. It is proving a huge success, at which I am not surprised, as I was trained by the RE myself in my school army unit. As a related anecdote in seeking professional help, when my son was himself at school and ran in the Cambridgeshire Athletics Championships in eastern England, I met another parent who attended the event. He is an Olympic gold medalist, and he told me that he and another gold medalist had formed a management consulting company which offered to bring all the qualities that successful sportspeople have to the operations of their clients.

As well as being better-prepared and better-informed to be in "Yes!" modality, one also needs to be resolute in saying "No!" when the occasion requires it. An interesting example of this befell me regarding an INGO Head of Office position for which I once applied in Sierra Leone. I attended the interview in London, sat a test in accounting and was offered the job. That day I was unable to meet the INGO Executive Director, as she was at another meeting. She was so respected a person, giving an excellent account of herself in television interviews, and I really did want to meet her, so I sought an appointment to do so. I was given one, traveled to London again at my expense and checked in at the organization's reception. I was told I needed to sit another accounting test. I replied that I had come not for that, but rather to meet the Director. I had already been offered the job and I was not prepared to sit another test. I made for the exit, whereupon pandemonium broke out … ah, it was OK, I didn't need to sit another test after all and the Director had invited me to her office straightaway. Up the stairs I went. The Director said, "I wanted to see if you could say 'No'". "I just have", I replied. We gelled immediately.

REFERENCE

Cervigni, R., Rogers, J.A., Henrion, M. (Eds.), 2013. Towards Climate-Resilient Development in Nigeria. Directions in Development: Countries and Regions. World Bank, Washington, DC.

CHAPTER 8

Importance of Local Knowledge in Building Resilience

Contents

8.1 Introduction	125
8.2 The "Groundnut Scheme" in Tanganyika	131
8.3 The Arctic Inuit	131
8.4 Sacred Sites in Liberia	133
8.5 Barking Dogs in Eritrea	134
8.6 Feedback on a Project in Central Asia	134
8.7 Crop Improvement Through Selection and Application	135
8.8 Harnessing Condensation for Drip Irrigation	137
8.9 Combining the Best of the Old and the New	138
8.9.1 The Best of the Old to the New	138
8.9.2 The Best of the New to the Old	142
References	144

8.1 INTRODUCTION

This may be a self-evident concept, yet it has many facets and its importance cannot be overestimated. Being "self-evident" has been realized only "recently" by "outsiders". Awareness of "local knowledge" has implications for "ways of doing things", and therefore relates to each of the OECD-DAC criteria which are used in any interim or final evaluation of an intervention (relevance, efficiency, effectiveness, impact, sustainability). Local knowledge also conditions the personal security of those involved in a development or humanitarian intervention at a given site.

Development has too often failed to deliver on its promises to poor nations. The policies imposed from above by international agencies and States have frequently not met the needs and aspirations of ordinary people. Some development agencies in their search for alternative approaches have pioneered "indigenous knowledge".[41] Yet how should such local knowledge be defined and who should define it?

In Chapter 1 of his book "Anthropology of Food", Pottier (1999) makes the case for anthropology being relevant to "better" development, through contributing to the thinking behind it and in practice. Anthropology must both clarify and challenge, he says, and he believes that food and agriculture is a suitable arena in which it can flourish. Agricultural research is a suitable "case for treatment", to ensure that it is conducted participatively, and that the objectives are relevant, community-based and sound. Technical progress as seen by an office-bound scientist or foreigner may not be perceived as such by the farming or fishing community, and it is through discussions at grass-roots level that farm or fish production/productivity initiatives should be designed and planned, building on initiatives that the community may already have in train.

It is true that anthropological research has revealed a great deal to foreigners who have enjoined the search for solutions to food insecurity. Pottier's comment on pages 39–40 of his book made me smile though, in relation to the asymmetries between men and women in access to agricultural and other resources, when he says that understanding of such can only be grasped through "detailed empirical research". Another way to achieve this understanding is for the foreigner concerned to go and live among these people long-term as one of them.[42] That is the route I have taken; my wife and I farm in Uganda. We don't need clipboards. Even when we are not in the country, not a day goes by when we are not in contact with our family there, hearing the latest gossip and happenings in our village. Perhaps this may constitute "detailed empirical research". In like manner, a foreigner who comes to live in England, say, will gradually and iteratively learn the customs and socio-economics of the people who live there, something new each day. This is arguably the better way, as the appearance of a clipboard distorts behavior of those being "studied", and a questionnaire can (unfortunately) be prepared and answers interpreted in such a way as to be nonsense in the given context (Pacey and Payne, 1985; Novellino, 2003).

[41]The term "indigenous knowledge" is not acceptable in some countries (China and South Africa, for instance), and elsewhere in this current book the term "local knowledge" is normally used.

[42]Rainer Maria Rilke wrote a remarkable piece of prose: "For the Sake of a Single Poem". When, 30 years ago, I came across a translation from the German by Stephen Mitchell, it was a telling masterclass for me as both a development worker and aspiring poet, for only those who follow Rilke's entry point and truly live with their "subject" for a long time will be able to understand the context, and contribute to it as he or she would wish.

Of course, I realize that not everybody can relocate as I did, in the way that I did, though not living there all the time owing to the demands of my consulting career. My wife has been my main informant about village life, and its complexities and intrigues, yet I have been one of her main informants too, seeing things that she does not see from my more objective perspective, in the very village in which she was born.

Relocating to India is what Belgian development economist Jean Drèze did some 40 years ago; he has written brilliantly on the challenges there, and how they can only be overcome if technical development, as measured by traditional development indices, is accompanied by social and ethical development (see Section 4.1).

There is no mention of resilience against food and nutrition insecurity in the index of Pottier's book "The Anthropology of Food", as it was published in 1999, before that term joined the lexicon of food security. Yet so much of what Pottier has to say does indeed relate to resilience, of key importance being the local perceptions, be they in rural or urban contexts, "as socially constructed, contested and negotiated", to use anthropological parlance (his page 16). He rightly acknowledges that food security and the ability to withstand famine are best understood as components of livelihood security (see Chapter 3, para. 1).

In the 1980s, social anthropologist "outsiders" came to appreciate that local knowledge had a value in "development" matters (Pottier et al., 2003). Publications like Robert Chambers et al. "Farmer First" in 1989 provided a corrective to the common outsider assumption until then that traditional beliefs and practices were obstacles to progress! This local knowledge came to be understood as an ever-evolving dynamic body of knowledge, subject to empirical improvement using ideas reaching the community from outside, and that knowledge production is embedded in social and cultural processes, this involving "social struggle, conflict and negotiation". This knowledge cannot be separated from the social context, and power relations within that, in which it develops. This includes the economic and political dimensions of the emergence and use of knowledge.

An example of this is from Bwisha in Zaire in which a husband and wife argue about whether the pernicious grass weed *Digitaria abysinica* is controlled by planting cassava. The wife claims it is, while the husband disagrees. The unstated subtext of the husband's contested position, however, is not the veracity or otherwise of the claim, but that he does not want his wife to plant cassava, as it is a crop that she controls, as per the local social norms (Fairhead, 1992).

Sillitoe (2002), in a book on "indigenous knowledge", makes his case for the involvement of anthropology in "development".[43] He does so not as postproject critic, pointing out how a given intervention should have been done differently, but as a key implementing partner contributing to positive change. By this he implies involvement in the whole development process from program conception and design through the whole project cycle. He laments the fact that there are others who are encroaching on anthropology's patch—agricultural economists and human geographers, even (horror of horrors) foresters and plant pathologists. I wonder what banana pathologists might have to say on that score (Tinzaara et al., 2013)? Sillitoe fears that these "others" may unwittingly sell anthropology short.[44] He gives short shrift to practices such as Rapid Rural Appraisal (RRA) and Participatory Rural Appraisal (PRA), both of which he rejects as "glib methodologies".

Sillitoe's working definition of "indigenous knowledge" in a development context is that it relates to any knowledge held more or less collectively by a population, informing understanding of the world. Ten characteristics of indigenous knowledge have been listed by Ellen and Harris (2000); these are in the context of environmental resource management, yet seemingly having more generic application. The three characteristics cited at the top of that list are

indigenous knowledge is local, rooted to a particular place and set of experiences, generated by people living in those places; it is orally transmitted, or transmitted through imitation and demonstration; and, it is the consequence of practical engagement in everyday life, constantly reinforced by experience, trial and error and deliberate experiment.

Sillitoe (2002) goes on to say that indigenous knowledge sits comfortably with a lot of anthropological research, anthropology having largely concerned itself thus far with the documentation and understanding of sociocultural traditions worldwide, which encompasses local knowledge by default. He opines that *research in anthropology* has thus far largely been an academic pursuit, its objective being to promote further understanding of the human condition, ultimately to elucidate how our sociocultural and biological heritage contribute to our uniqueness among animals. *Research in local knowledge*, on the other hand, he says, relates to development issues and problems, its objective being to introduce a locally informed

---

[43]"Participating in Development" was the title of the Association of Social Anthropologists (ASA) of United Kingdom and the Commonwealth Conference in London in 2000, from which this volume by Sillitoe (2002) and its two companion volumes (*Development and Local Knowledge* and *Negotiating Local Knowledge*) were derived.

[44]For plant pathologists, foresters and other development workers to imbibe and apply some of the social skills of a sensitive "radical ethnographer" (see Section 4.2.1.7) in their field work is surely to be applauded.

perspective into development, to challenge the assumption that development is something outsiders have a right to impose, and to promote an appreciation of indigenous power structures and know-how.

There are methodological challenges relating to the transition from academia into development though, with Schönhuth (2002) cautioning the inconsistencies between participatory and academic anthropological research, and that some anthropologists may feel their intellectual integrity threatened by anthropology taking this "exciting" new direction. Schönhuth voices two anthropological reservations towards "participatory approaches", namely (a) in their efforts to produce timely and action-oriented results, approaches such as RRA and PRA fade out crucial parts of local reality as well as the sociocultural dimension, and (b) participatory approaches contain implicit or explicit assumptions that relate to Western norms and discourse, but do not relate to the cognitive structures and decision-making processes of local cultures.

Sillitoe (2002, page 11 in his Chapter 1) continues:

> In allowing local populations to inform the development process with their own knowledge and aspirations, we open up the prospect ultimately of a redefinition of the meaning of development itself and its aims. This is where the indigenous knowledge in (the) development agenda inexorably leads, although until now we have avoided speaking of it, anxious at the reaction of international funding agencies. It represents the reduction of foreign hegemony, the promotion of what some call endogenous development. Development agencies are likely to see it as subversive because it will diminish their control. When we interpret indigenous knowledge in this wide sense, we have local stakeholders contributing equally to the determination of development objectives alongside foreigners mindful of their funding agencies' policy initiatives, reducing the control of the latter, whether or not disguised by participation rhetoric.[45]

Professor Sillitoe entreats (his page 20) anthropologists to "work more with other disciplines ... not in isolation as previously ... with the noble aim of affording a voice to the previously marginalized and post-colonized, notably helping the very poor".[46] In his view (page 18), anthropology faces

---

[45]See Section 6.2.5.2 regarding the superb "participation" within the *Qali Warma* program, Peru, and that involved in formulating the Palestine National Food Security Strategy (Section 7.2).

[46]I strongly endorse this sentiment. As a trained psychologist I do not regard economist Richard Thaler's excursion into "my" subject as trespass, the added value resulting in his being awarded the Nobel Prize for Economics in 2017 (Section 4.1)! The overlap of social anthropology, sociology, psychology, agriculture, economics, archeology and education (to name but a few disciplines) is surely becoming ever more apparent, meaning that the disciplines' adherents will listen to, and negotiate with, each other the more, which has to be a good thing.

the danger of lack of intellectual identity or clear purpose, as the discipline's core subject matter, namely small-scale largely independent societies, finally disappears at the start of the 21st century under the relentless pressure of colonization and subsequent globalization. Indeed, the whole volume in which his paper appears, in Sillitoe's words (page 14), "seeks to advance needed theoretical and methodological debate, to address ontological and epistemological issues, and to put this vital (indigenous knowledge) research on firmer intellectual foundations to rebuff critics and justify its place in development practice".

Other contributors to the Sillitoe, Bicker and Pottier volume of 2002 cited in the references are similarly passionate and convincing, none more so than Darrell Posey (2002) in Chapter 2 who encourages "a new kind of anthropology", saying (pages 39–40):

> with these changes will emerge a much more dynamic, responsive, and relevant anthropology. It will be an anthropology that guides the multidisciplinary and multicultural responses that are needed to solve global problems, while respecting what indigenous peoples call "the sacred balance".

Using local knowledge in program design and planning has the advantage of rendering that program more likely to be technically successful and long-lasting, thereby building resilience and sustainable benefits, especially if the program builds on past or existing already-appreciated development efforts. Furthermore, and at least as important, is the sense of ownership by the local community, when a development project is predicated on its indigenous knowledge and needs, and that community is involved in determining the project's objectives and participates in its design, planning and implementation. During this process, local capacity and confidence is built, technically and in terms of aspects like timing and sequencing, and indeed capacity built of any outsiders involved. "Insiders" and "outsiders" will reinforce each other, in a win-win situation, in which local culture can be both preserved and strengthened, while socio-economic conditions can be improved equitably. The local community will certainly help in the internal communications requirement so vital for a program's success, including the need for technologies from outside to be adapted to suit local conditions and needs, as necessary.

It is "better late than never" for local knowledge to be appreciated, often of the poorest and most dispossessed people on the planet, being recognized by outsiders as being instrumental in improving global wellbeing, through its incorporation into development which is sustainable and environmentally friendly. Yet "The richness of indigenous knowledge is in the doing not the telling" (Croal and Darou, 2002).

An example of "the doing" is the "Common Narrative" exercise, which was recently undertaken in Bangladesh (facilitated by the SUN/ UN Network for SUN/REACH). The aim of this initiative was to bring together the main stakeholders for nutrition to work jointly in presenting a Common Narrative on undernutrition, based on joint analysis and interpretation of the current nutrition situation. It can therefore provide an important starting point and be used as a basis for discussions at various levels and across sectors, and with a range of stakeholders at local level. Ultimately, the purpose of the Common Narrative is to assist the government, development partners, civil society and others in understanding clear policy and program goals, and monitoring progress towards common targets. The Common Narrative is generally thought to have played an important role in building consensus and awareness within the ongoing process of dialogue; this is as important as, if not more so, than the short document which came out as a result of the initiative. The initiative successfully generated and factored in local knowledge, which was instrumental to the project's success.

Examples are given below, all but two from the author's own experience, of how such local knowledge can facilitate, and lack of it hinder, the development of resilience against food and nutrition insecurity.

8.2 THE "GROUNDNUT SCHEME" IN TANGANYIKA

The British government's "groundnut scheme" in Tanganyika (Tanzania) in the late 1940s failed because the soils were too clayey and the rains insufficient to enable crops to mature. The scheme was imposed by outsiders, and local indigenous knowledge was not sought. The people of the area, who knew well the local edaphic conditions, would surely have counseled against the scheme, had their views been sought (Fuglesang, 1984). The result was a costly, wasteful investment and a dustbowl created where once there was dryland bush protecting the soil surface from the elements.

8.3 THE ARCTIC INUIT

The traditional Inuit lifestyle of hunting and gathering is fast being replaced by life in settlements, with access to modern facilities and foodstuffs (see also Section 11.2, Case Study 2). However, settlers still require their traditional foods to some degree to maintain their identity and thereby their cultural resilience. The gender demographic of the traditional Inuit shows that there are more females than males, one reason being that it is the men who do most of the hunting, which is a highly dangerous lifestyle under Arctic conditions.

Local knowledge is crucial for these hunters, to navigate their way around the frozen environment, and locate their quarry, kill it and bring the bounty home to their families. A graphic account of the danger is given on an Inuit website, recorded by a non-Inuit (Glenn Morris)[47] who was accompanying a hunting party during an expedition he led to the Canadian Arctic in 2007–09 (Box 8.1).[48]

BOX 8.1 Hanging by a Thread in Greenland

I was wrapped in caribou fur and was wearing polar bear skin trousers, still the wind stung my face as my host Lars Jeremiassen and I sledged on the frozen sea off the coast of Steensby Land. I had rigged up a rope trapeze and suspended myself from the back of the sledge to allow me to take video footage of Lars and the dogs as they trotted along. With no warning Lars sent the whip hissing through the air and with a sharp command the dogs leapt forward into a fast run. I just about managed to hold on and laughed, explaining in my video commentary that Lars was showing off. On our return to Qaanaaq a month or so later Lars' daughter who, unlike Lars, spoke English, asked if I knew why Lars had whipped the dogs into a run. I said that I thought he was having a bit of fun—"No", she said, "the ice was breaking up under you".

I often tell this story when giving talks or presentations as it seems to me to illustrate an exceedingly important point. The Inuit have lived in the lands of the north for thousands of years. During this time they have built up a respect and understanding of their surroundings and environment that is both deep and intuitive. They know and connect with their land in a way that we, in the industrialised and wealthy south, have all but lost. To me, the noise of the sledge runners squealing and clattering over the ice sounded normal, whereas Lars heard the difference between life and death.[49]

---

[47]Glenn Morris is a British explorer and founder of Arctic Voice, an organization dedicated to highlighting the impact of climate change on the Arctic and its people.

[48]http://www.arcticvoice.org/inuit.html (accessed August 29, 2015) (quoted with permission).

[49]http://www.wayne.k12.ga.us/news.cfm?story=51275&school=685 (accessed August 29, 2015) (quoted with permission).

8.4 SACRED SITES IN LIBERIA

Just as it is important for the Inuit to maintain strong links with their history and traditional lifestyle, so it is important for Liberians to do so. The country was ravaged by civil war (1989–96 and 1999–2003) and there was death and destruction across the country, and massive displacement of people. In 2011 the current author visited the ArcelorMittal mining site in Nimba County near the northern border with Guinea and Ivory Coast, and was shocked to come across the burnt-out remains of heavy mining equipment, their huge tires torched by a militia group at the time which destroyed the whole site. He also met James Jallah, long-time Yekepa resident who holds the institutional memory of the area (see Section 5.7). Jallah also serves as cultural heritage assistant/camp coordinator of ArcelorMittal Liberia's community liaison department. He told me that many of the youth there have no knowledge of their cultural history, or the artefacts created by their ancestors.

Now that peace has been restored, the mining operation is set to expand, and provide livelihoods and food security to many local people in a community still in shock at the devastation of their country so recently. To its credit, the company has a Cultural Heritage Program managed by a social anthropologist, which forms part of the company's Resettlement Program for villagers who will be displaced by future mining operations. An account of part of its work can be found on the web, from which the following account is extracted.[50]

While planning for the start of the Direct Shipping Ore (DSO) phase of its mining operations, ArcelorMittal Liberia conducted surveys in the villages close to its mining sites at Tokadeh, Yuelliton and Gangra to determine the location of sites that are sacred to these communities. Consultative meetings were held with cultural leaders and residents of the villages surrounding the proposed mining sites, and they identified several sites that they revered and would want to see protected for posterity.

Six sacred sites have been relocated while two sites have been fenced *in situ*. The Golodien Rocks, situated in the village of Golodien, near Tokadeh, is one of the sites which were fenced. The rocks are protected not just by a fence but by a national Cultural Heritage Preservation Order, which prohibits people from destroying the site. The other preservation order covers the Bontor Family Graveyard near Gbapa. The protected status enables the community to remember their great chiefs, heroes and warriors in perpetuity.

[50]http://www.nanews.net/news/liberia-200-year-old-sacred-rocks-preserved/ Nordic Africa News website, April 16, 2015 (accessed August 29, 2015).

The Golodien Rocks were earmarked as important to the community in 2008, after they were deemed to be potentially threatened by works to develop Phase 2 of the company's Liberia operations, as they were located in the proposed area for the company's concentrator plant. The Rocks are important historical and cultural artifacts for the residents of Golodien and peoples extending as far as neighboring Guinea. The village is situated on lands owned by the late great Pa Golodien. At his passing, over 200 years ago, his family decided to build a monument in his honor with rocks brought from the original family site in present-day Guinea.

This cultural awareness program builds empathy between the mining operation and the people, which facilitates industrial relations and helps create wealth for the people in this remote area, enabling them to purchase the food they need from the shops and the small farmer's market. The program thereby builds resilience in the community.

8.5 BARKING DOGS IN ERITREA

I tell this story against myself, showing my own ignorance as a visiting "expert". In 2001, I visited Eritrea on a short-term project formulation mission, in the Gash Barka depression bordering Ethiopia. The area was contaminated by landmines from the then-recent border war. I had not visited the area previously. I met farmers who grew bananas under tube well irrigation. One lamented that herds of elephants sometimes came, attracted primarily by the irrigation system as a fresh water supply. They destroyed not only his plantation but also the pumping equipment and pipes, which he then had to repair. Trying to be helpful, I suggested tethering a dog near the equipment, in the hope that its barking would deter the herd. My suggestion was met with scorn. The farmer pointed out that the elephants would not be deterred—on the contrary, they would rush the dog which would break its tether in panic and flee to its master's house, with the elephants in hot pursuit, which would then trash the home and run amok in the banana plantations. Clearly, the farmer spoke with the experience of this chain of events, perhaps from having tried that remedy before. My proposed "solution" would have done nothing to bolster resilience of the area against food insecurity!

8.6 FEEDBACK ON A PROJECT IN CENTRAL ASIA

While working as a livelihoods adviser to a government Minister in Central Asia, I was invited to attend a presentation by a university research team on the rural economy of a particular community. This research had been undertaken at considerable expense, funded by a European government, through an association of a European university department of international development and a local research center. I asked my team of nationals from within my ministry to accompany me, so I could benefit from their perspectives on what we were to hear on the livelihood strategies of the community concerned.

After the presentation I asked my team one by one what they had learned from the graphs and pie charts of the PowerPoint presentation, and the analysis of village livelihoods provided by the foreign researchers. To my surprise, the answer was "Nothing". They already knew the portfolio of livelihoods followed by their rural communities to derive their income. They had lived the life described by the researchers for 20 years before they left home to attend university in the capital. They knew better than the researchers ever would the livelihood strategies that their parents and neighbors followed.

The research undertaken had therefore benefitted the foreign university staff, because they were not previously aware of the dynamics of village life in that country, and the donor concerned, who would receive their report. The research could be written up as a paper and published in a learned journal, to help secure their own tenure and livelihoods at their university. Yet the key information gathered from the research project could probably have been gleaned by a simple PRA over tea and biscuits with opinion leaders in that village far more cost-effectively and quickly, and used to inform a development initiative to enhance ongoing indigenous initiatives. When I explained this to the head of section in the donor agency concerned, the realization shocked him.[51]

8.7 CROP IMPROVEMENT THROUGH SELECTION AND APPLICATION

An example from Africa is the benniseed (sesame) being grown in Katsina State and elsewhere across northern Nigeria. The fact that the fruits (capsules) shatter is regarded as the "norm", just as diarrhea in infants is regarded as the "norm" by women who visit the CMAM nutrition clinics there for the first time. Neither should be the norm. In the case of benniseed, there are "cultivars" which do not shatter, which

---

[51]Both the detailed research and the PRA would have missed key anthropological and sociocultural factors, as lamented by both Sillitoe and Schönhuth (Section 8.1).

could be introduced and tested. In Gedaref State in Sudan, for example, nonshattering sesame is the norm, as I realized when working there in 2016. Importing and planting seed of such in Katsina State, by farmers who were previously unaware that there are types of benniseed that do not shatter, could mean that much of the yield would no longer spill on the soil before or during harvesting operations, and onto the compound floor when the harvested plants are stacked to dry around the homestead awaiting threshing.

Walking around Daga village, Katsina, on September 30, 2014 I observed to my Nigerian colleague, a professor of nutrition, that the chickens which ate the spilled seed were perhaps some of the best-fed village chickens in Nigeria, sesame being a rich source of protein and oil. All that is needed is a PVS trial in which at least one nonshattering variety is tested against the local control. For this to be selected by the farmers concerned, its other characteristics combined, apart from net seed yield, must be deemed no worse than the farmers' control types. It is for an agency such as the State government Department of Agriculture, a Consultative Group on International Agricultural Research (CGIAR) body, or a project actually to cause this to happen. Application is vital. Lobbying is needed. Selection of one or more cultivars by breeders or biochemists alone is insufficient, for the key issue downstream is whether the farming community endorses one of their selections.

The need for local champions, who could promote a biofortification initiative in major food crops in the north of the country, is mentioned in Chapter 10. The best technical lobbyist engaging with the nonshattering benniseed opportunity could well be a nutritional anthropologist, familiar with the local culture and language, working closely with local social structures and their champions. Nutritional anthropology is a "biocultural paradigm" that sets out to explain how social structures and actions interact with ecological and biological systems (such as agricultural practice, food processing and cultural norms) to determine food availability and use, and nutritional health (Chrzan, 2013). Janet Chrzan's paper looks *inter alia* at how biological and cultural variation affects dietary practices and health, and how food use and nutrition interacts with parasitic infectious and chronic diseases.

Also with benniseed, a huge range of protein content in cultivars grown in Nigeria has been shown by in-country research. The values of 30 Nigerian cultivars examined in a local study ranged from 3.25% to 11.27% (Nweke et al., 2011). Only 11 of the 30 cultivars had more than 7.26%

protein content. There is scope for ensuring that within the nonshatter-ing genotypes selected in breeding and selection trials in Nigeria, there is also selection for high protein. Any correlation between these two variables among the 30 cultivars tested is not revealed by Nweke et al. The seed of benniseed is rich in "limiting amino acids", which determine the rate of protein synthesis in the human body, and also in calcium, necessary for bone development and repair. The protein content of benniseed worldwide can be as high as 28%, so there is some way to go for breeders in Nigeria to ensure that protein content of this crop is raised.

Local knowledge need not be the preserve of local people, however. In the case of Nepal in 1996–98, for instance, an intervention was based on knowledge that I had, but the local people did not. I was at the time working in a development project in Arghakhanchi district in the *west* of the country. Much of my work involved trying to improve the "variety" of some of the key crops grown by farmers in the low hills of the Himalayas, up to 1500 m, using the PVS methodology, as described in my earlier book of 2016. One of these crops was oilseed rape, the main source of cooking oil for the villagers. It happened that I had previously worked in the same agro-ecological zone in *eastern* Nepal for several years. While there, I had come across a "variety" of rape being grown around the village of Hile in Dankhuta district. It grew and yielded very well, and I thought it might perform equally well in Arghakhanchi. And it did, receiving rave reviews from the farming community in terms of yield, resistance to both disease and lodging, and taste. There was no need for any extension advisories or leaflets to be distributed. Its popularity was assured by referral, to relatives and neighbors.

8.8 HARNESSING CONDENSATION FOR DRIP IRRIGATION

In my first year at college I was fortunate enough to have an inspiring lecturer who told me "things that you will not find in textbooks". One of those "things" was that there is more water in a unit volume of hot desert air at midday than there is at the same time of day in the same unit volume of air on a foggy autumn day in Britain. This is counter-intuitive, until one experiences it. Fast forward a few years, and I found myself in Libya sleep-ing on a Mediterranean beach under the stars, finding I was drenched with water by daybreak. The only dry bit was the pillow beneath my head. As the temperature drops at night, the moisture in the warm air condenses onto everything it can reach. This is the source of water for plants in the wild and crops, condensation collecting on both above-ground parts and the roots below the soil. Desert insects survive on water that condenses and coalesces as droplets on their outer surfaces (cuticle); these droplets run down to their mouths when they angle their bodies semivertically.

During the Roman occupation of north Africa, this condensation principle was used to irrigate grapevines, through farmers arranging rounded stones from wadi beds around the base of those vines. I mentioned in my earlier book the grasses which grew along the tarmacked road in Sebha in the Libyan Sahara, watered by the run-off of condensation during the night. Likewise, the metal surfaces of vehicles in the desert are covered in droplets which have condensed there overnight.

Both irrigation and domestic water can be harvested free from nature in this way by using a surface which will serve to trap such liquid water from the cooling night air. Hanging nets vertically is one way to do this (the reader may wish to check out "fog harvesting" and "cloud catchers" on http://www.youtube.com, in Peru, Chile, India and elsewhere).

8.9 COMBINING THE BEST OF THE OLD AND THE NEW

8.9.1 The Best of the Old to the New

At an agricultural extension conference I called at Makurdi, Nigeria in 1999, the star attraction was a Nigerian octogenarian former extension worker. He told the assembly how extension used to be conducted in the colonial era, on horseback to the most remote areas, and it really did work. The audience mused on how remote from the farmers' real concerns the modern extension system had become when compared with the personal contact established then. The way to bring about the marriage of the best of what is old to the best of what is new had still to be devised in 1999 (see also Section 11.1.2.1, point B). Photos 8.1 and 8.2 show how old technologies can stand tall alongside the new.

Another colonial relic of technology which could bring the best of the old back to the fore was one that I came across at Suam on the Uganda side of the border with Kenya (Trans-Nzoia county and the Kitale highlands). An old colonial farm there, Kapyoyon Farm, was now in the hands of a local cooperative, and I stayed there when I was on mission to Mount Elgon from 1991 to 1994. It was the only part of the country which was sufficiently high and therefore cool to support good crops of barley. While there, I saw the old hydraulic ram pump technology which brought water from the perennial Suam river (stream) up to the farm, a system which was

Photo 8.1 Traditional and more modern 4WDs at Hamish Koreib market, Kassala State, Sudan (April 29, 2016).

Photo 8.2 Rehabilitated Roman-occupation era water cistern at Frube village near Al-Mazar Al-Shamali, Irbid governorate, NW Jordan (July 13, 2016).

then in disrepair. The technology is so simple, with water forced along a pipe of decreasing diameter, which created sufficient water pressure to raise it up the hill to the farm, with no other power source involved except the kinetic energy of the stream flow.

Modern-day modifications of the old design specifications render this technology even more relevant to the increased need for delivery of domestic water to remote places, in a sustainable way, and it could be set for something of a renaissance—recycled local knowledge of a previous era. Human resilience is so dependent on stable supplies of food *and* domestic water; the stream can also provide sustainable electrical energy, using a water-powered generator suspended in it, and of course agricultural water, which can be raised by a paddle wheel suspended in the stream flow to lift it into irrigation canals on the banks.

I came across another instance of indigenous knowledge of "old technology" in Aohan, Inner Mongolia, it not being properly appreciated by national policy makers until the intervention of Martin Jones, Professor of Anthropology and Archaeology from the University of Cambridge, United Kingdom.[52] This involved a group of small grained cereals, the millets, very much the country cousins of the large-grained cereals, corn, wheat and rice, which comprise the bulk of global cereal grain trade. It was Professor Jones' conversations with local farmers that radically altered his perception of the grains. Clearly these farmers, just like their ancestors over thousands of years, had needed resilient plants that could ripen to harvest in challenging years, to ensure food security for the population (Box 8.2).

BOX 8.2 The View From Inner Mongolia
Professor Jones says:

When we first visited Aohan it could sometimes be hard to tell whether the millet was growing as a crop or as a weed. We asked the locals, and rather than tell us it was a stupid question—that it was irrelevant whether it was crop or weed—they politely answered a different one. They told us what it tasted like and when they last ate it. These people had lived through hard times, famines, so to survive they had developed more open ideas. I realised then that I'd come with concepts that seemed universal but just weren't relevant to the lives of people in contemporary northern China.

---

[52]http://www.cam.ac.uk/research/features/archaeology-shows-theres-more-to-millet-than-birdseed (quoted with permission).

BOX 8.2 The View From Inner Mongolia *(Cont'd)*

Jones was aware that millets retain the genetic characteristics which convey resilience to drought, traits which were lost during the breeding of corn, for example, through drought-avoiding short maturity period and other physical and physiological features.

Being geared towards producing heads of large grains is terrific if you can guarantee all the water, nutrients and sunlight they need. But the crops are much more prone to complete failure if something changes, like the amount of rainfall in a growing season. For farming systems where there's no financial infrastructure providing subsidies and grants to help farmers control the growing conditions through irrigation, pesticides and other methods, inherent crop resilience can be vital to a successful harvest.

The Aohan government has responded very positively to the suggested need for a focus on the traditional small-grained cereal crops, which Jones and his Chinese counterparts have recommended. Furthermore, the FAO has recognized the Aohan Dryland Farming System as a "Globally Important Agricultural Heritage Systems" site.[53]

Other contributions to the longevity and resilience of current traditional agricultural systems are being pursued by archeologists and anthropologists, under the banner of "applied archeology"—for instance, how long a particular agricultural system visible today has been going on.[54] To mention just one example of each of the disciplines involved, I cite Daryl Stump, archeologist with the University of York, United Kingdom and Amanda Logan, anthropologist with Northwestern University, Evanston, Illinois, United States (Stump, 2013; Logan, 2016). Stump is examining the long-term sustainability of two East African agricultural systems, at Engaruka in Tanzania and Konso in southwest Ethiopia. His 2013 article is a review of work done on the relationship between archeological knowledge and indigenous knowledge.

[53]In some dryland areas there are sociological reasons why small-grain crops were replaced by large-grain—sorghum by maize in the Ukambani area of Kenya, for example. In this case it was that as schooling became more available, children were no longer available to serve as bird scarers in the fields, and small-grain crops were decimated (see Harrison, M., 1970. Maize improvement in East Africa. In: Leakey, C.L.A. (ed.), Crop improvement in East Africa. Farnham Royal, United Kingdom, Commonwealth Agricultural Bureaux.

[54]The example of the Pharaonic Nile Valley was given in the current author's earlier book (Chapter 3.3), where contemporary stelae can be read for what they tell us of droughts in ancient times and how governance and society were changed for ever because of them.

Logan addresses how the problem of hunger came to take its present-day form. Combining archeobotanical, ethnoarcheological and environmental data, she shows how food insecurity was avoided during a centuries-long drought in Banda, Ghana. Such studies are useful for areas of the world that are lacking rich written archives. She strongly advocates the need to historicize food insecurity. Her thesis is that the diverse economic and social strategies used by people to survive drought in the past have been eroded by centuries of economic and political exploitation, reducing their optimizing strategies. She characterizes this as an example of "slow violence". Now, even minor environmental shifts have major impacts on food security.

8.9.2 The Best of the New to the Old

Agricultural research in developing countries has hitherto often been conducted in a top-down way, rather than being led by the priorities the farmers have. Getting research and extension better linked, and both centered on-farm, remains a priority. Even without the presence of college-qualified specialists, farmers conduct their own "experiments", having done so since the start of settled agriculture (Okali et al., 1994; Sumberg and Okali, 1997).

Okali et al. (1994) discuss participation of farmers in their own and "official" research, empowering farmers, the importance of local knowledge systems and the role of NGOs. Their book considers how best to bring together the way farmers do research and how a farming systems researcher might want to do it. The value of qualitative experience is shown to be highly important. Sumberg and Okali (1997) also relate the need to involve farmers in "official" agricultural research, with many examples from the field. Experimentation in the farming community in developing countries is widespread, yet overlooked and undervalued by "official" agricultural researchers. The authors investigate the potential for a better integration of official research with this indigenous research, reviewing the literature on the same and conducting their own fieldwork in Africa. They reviewed the characteristics of those who experiment in the informal system and those who are at the center of local information networks. The latter group are largely opinion leaders, respected by their neighbors, often men who hold public office and/or have some formal technical training and a connection with formal research and extension services. Other sources of influence on villagers include private sector suppliers of agricultural/veterinary inputs and NGO personnel who bring development projects into their midst.

Another example of bringing the "new to the old" is from Jordan. There are 2,000 farmers in the Ghour al Safi region of the Jordan Valley, on the Jordanian side. Each farm has a raised reservoir (pond) lined with heavy-duty polythene to receive water which is fed to it from the Al-Mujib dam and some private and public groundwater wells in the highlands, every three days by the Jordan Valley Authority, to be used for irrigation. The Al-Mujib dam main supply comes from water streams and the run off that is harvested from the huge catchment area around it, so the water is clean and of good quality. Each of these ponds has the potential to stock fish to supply the current insatiable market for fish in the country. There are other regions of the Jordan Valley that also have such irrigation ponds. Fish fingerlings are available from private and government sources, and imported fish feed pellets from private sector suppliers.

Many smallholder farmers have responded to government of Jordan initiatives and the aquaculture component of the part-DANIDA-funded Middle East Regional Agricultural Program (MERAP) to stock their irrigation ponds with fish, thereby providing an income stream additional to that from their horticulture and ruminant enterprises. In June 2016, the current author visited a number of these fish farms, government research stations, and private suppliers.

Abdullah Abu She'reh is one such fish farmer in south Shuna. Under MERAP Phase III, he procured 1,000 fingerlings, both tilapia and carp, and is using his irrigation pond to raise them. He was one of many farmers under the program who responded to an extension campaign, offering 500–1,000 fingerlings free of charge; these fish mature and become their own brood stock. Such is this farmer's enthusiasm for aquaculture that he is constructing other ponds on his land (Photo 8.3).

Photo 8.3 Fish feeding frenzy in response to concentrate pellets being thrown into an "irrigation pond", lower Jordan Valley, Jordan (July 11, 2016).

Photo 8.4 Fish fed on bread only, in low-input low-output modality, mid-Jordan Valley, Jordan (July 10, 2016).

Hamed Yusef Al Kareer is another "new" fish farmer at Dier'alla. This farmer has chosen to spend nothing on imported balanced ration feed, instead supplementing the fish diet of grazed algae off the polythene lining and what is free floating in the pond with old pitta bread as it becomes available from his household. The growth rate is far lower than that of fish in Mr. She'reh's reservoir, which have been better-nourished, but still Mr. Kareer has mature fish to sell, and a supply in perpetuity of his own home-grown fingerlings (Photo 8.4).

There are many indicators of success from this example of "bringing the best of what is new to the best of what is old". One is that the Jordanian government responded to the MERAP aquaculture initiative by creating an Aquaculture section in the National Center for Agricultural Research and Extension (NCARE); second, farmers seek out aquaculture extensionists now, which is the reverse of the situation when MERAP commenced its aquaculture intervention; third, the high demand for tilapia fingerlings, mature fish and fish feed concentrate from the Ziad Atalla commercial aqua-farm in the Valley at Karameh, South Shuna.

REFERENCES

Chrzan, J., 2013. Nutritional anthropology. In: Albala, K. (Ed.), Routledge International Handbook of Food Studies (2012). Routledge, Abingdon, pp. 48–64 (Chapter 5).

Croal, P., Darou, W., 2002. Canadian First Nations' experiences with international development. In: Sillitoe, P., Bicker, A., Pottier, J. (Eds.), Participating in Development: Approaches to Indigenous Knowledge. Routledge, London, p. 105 (Chapter 5).

Ellen, R., Harris, H., 2000. Introduction. In: Ellen, R., Parkes, P., Bicker, A. (Eds.), Indigenous Environmental Knowledge and Its Transformations. Harwood Academic, Amsterdam, pp. 1–33.

Fairhead, J., 1992. In: Indigenous technical knowledge and natural resources management in sub-Saharan Africa: a critical review. Paper Commissioned by the Social Science Council, New York.

Fuglesang, A., 1984. The myth of people's ignorance. In: Development Dialogue. Dag Hammarskjold Foundation, Uppsala, p. 51.

Logan, A.L., 2016. Why can't people feed themselves? Archaeology as alternative archive of food security in Banda, Ghana. Am. Anthropol. 118 (3), 508–524.

Novellino, D., 2003. From seduction to miscommunication: the confession and presentation of local knowledge in 'participatory development'. In: Pottier, J., et al. (Eds.), Negotiating Local Knowledge. Pluto Press, London. 277 ff. (Chapter 12).

Nweke, F.N., Ubi, B.E., Kunert, K.J., 2011. Determination of proximate composition and amino acid profile of Nigerian Sesame (*Sesamum indicum* L.) cultivars. Niger. J. Biotechnol. 23, 8–12.

Okali, C., Sumberg, J., Farrington, J., 1994. Farmer's Participatory Research: Rhetoric and Reality. Intermediate Technology Publications/ODI, London. 159 pp.

Pacey, A., Payne, P. (Eds.), 1985. Agricultural Development and Nutrition. Hutchinson/Westview Press, London/Boulder, CO.

Posey, D., 2002. Upsetting the sacred balance. In: Sillitoe, P., Bicker, A., Pottier, J. (Eds.), Participating in Development: Approaches to Indigenous Knowledge. Routledge, London, pp. 24–41 (Chapter 2).

Pottier, J., 1999. Anthropology of Food: The Social Dynamics of Food Security. Polity Press, Cambridge. 230 pp.

Pottier, J., Bicker, A., Sillitoe, P. (Eds.), 2003. Negotiating Local Knowledge: Identity, Power and Situated Practice in Development. Pluto Press, London, p. 332 (Chapter 1).

Schönhuth, M., 2002. Negotiating with knowledge at development interfaces. In: Sillitoe, P., Bicker, A., Pottier, J. (Eds.), Participating in Development: Approaches to Indigenous Knowledge. Routledge, London, pp. 139–161 (Chapter 7).

Sillitoe, P., 2002. Participant observation to participatory development: making anthropology work. In: Sillitoe, P., Bicker, A., Pottier, J. (Eds.), Participating in Development: Approaches to Indigenous Knowledge. Routledge, London, pp. 1–23 (Chapter 1).

Stump, D., 2013. On applied archaeology, indigenous knowledge, and the usable past. Curr. Anthropol. 54 (3), 268–298.

Sumberg, J., Okali, C., 1997. Farmers' Experiments: Creating Local Knowledge. Lynne Rienner Publishers, Boulder, CO. 186 pp.

Tinzaara, W., et al., 2013. Communication approaches for sustainable management of Banana Xanthomonas wilt in East and Central Africa. In: Blomme, G., van Asten, P., Vanlauwe, B. (Eds.), Banana Systems in the Humid Highlands of Sub-Saharan Africa: Enhancing Resilience and Productivity. CABI, London, pp. 224–234 (Chapter 27).

CHAPTER 9

Lateral Thinking

Contents

9.1 Introduction	147
9.2 Flamingo Breeding	149
9.3 SWOT Analysis	150
9.3.1 Strengths	150
9.3.2 Weaknesses	150
9.4 Population Management	155
9.5 Bringing a New Idea to a Community	155
9.5.1 A Vision for Eco/Agro-Tourism	155
9.5.2 Rapid-Impact Technical Interventions	157
9.5.3 Differing Perspectives	159
9.6 The Value of Corn Cobs in a Parched Field	161
9.7 Potato Promotion in France	161
9.8 Nepal Earthquake in 1998	163
References	165

9.1 INTRODUCTION

It is important for planners and implementers to keep their eyes and ears open for evidence in the community about its nature and needs, which may not be voiced or otherwise indicated for one reason or another, yet which may explain much about their condition with regard to their capability of building resilience against food insecurity.

Annex 1 comprises a 5,000-word composite essay which starts with food security and then takes on a life of its own to show how the mind can be agile if allowed to operate in a "lateral" divergent way (thinking outside of the box), rather than a convergent way (which is also an essential modality much of the time). There is room for both in life, including seeking to improve our resilience against food and nutrition insecurity. Allowing the mind to be agile, to operate in a "lateral" way some of the time enables peripheral thinking to enter the mainstream, bringing with it fresh ideas, new connections. Were I a "marketing" professional, I could perhaps summarize

Human Resilience Against Food Insecurity
https://doi.org/10.1016/B978-0-12-811052-2.00009-3

the essay as "a lateral-thinking stream of anthropological consciousness on the theme of food security!".

Regarding my earlier book, one of the peer reviewers of the initial plan criticized me for not saying there and then what my "conclusions" would be in the last chapter. The fact is I didn't know before I had written the rest of that book, any more than I know now, at this chapter of this book, what all of my conclusions will be in Chapter 12. Writing a book like this is an act of self-discovery, collating and analyzing what I have learned from my craft over almost five decades.

As author of that earlier book, I had no final idea of *all* that I would say when I started it—*much* of it yes, but not *all*. Similarly, as author of the essay in Annex 1, I had no idea what direction the essay would take when I started writing about goats. I let myself be influenced by the lateral thoughts which came into my mind as I contemplated the subjects I was addressing. That is how most of the poems I write develop too. I was pleased to note that Californian Poet Laureate and critic Dana Gioia (mentioned in the introductory Chapter 1) shares that view: "I rarely know where a poem is going until it is finished. In fact, if I know initially how a poem will end, I lose the impulse to write it".[55]

As part of our degrees in psychology, my fellow Cambridge student (P.D. Evans)[56] and I undertook some joint research by employing a series of tests to measure the degree of convergence or divergence in thinking and perception patterns that the University undergraduates had, across disciplines. This work was rooted in a book written in 1966 by Liam Hudson, called "Contrary Imaginations", which explored the thinking patterns of schoolboys as related to the tension between intelligence and creativity, and subsequent career choices. In a very small trial, we found that scientists tended to be convergent thinkers, whereas arts students and engineers were divergent thinkers. Divergent thinking can lead to unexpected serendipitous places, the journey potentially opening up new connections and pathways which

---

[55]http://danagioia.com/interviews/paradigms-lost-interview-with-gloria-brame (accessed July 5, 2017) (quoted with permission).

[56]Professor Evans' subsequent research at Westminster University has included investigating the effects of common stress situations in our lives on the body's infectious disease immunity. This line of research has application for the food-insecure, who of course are highly concerned about feeding themselves and providing for their families. The food-insecure are likely in double jeopardy, of being both nutritionally- and psychologically-more predisposed to succumbing to infectious disease.

could not be explored by a convergent thinker, who is all the time reducing options.

Lateral thinking requires keeping one's eyes open, sometimes looking slightly away from the object, as at night to get better resolution, or to use English colloquialisms, using eyes in the back of one's head or a third eye. A conversation comes to mind that I had in Moroto, in the remote and dry Karamoja district of Uganda, with a Ugandan helicopter pilot. He told me that he could tell which herd of cattle had been rustled by flying low over the animals and their herders—rustlers would scatter assuming they had been caught, whereas if the herders were the owners, they would stay with their livestock. Some other examples are cited below, comprising various categories of lateral (divergent) thinking, some directly related to food security, others only indirectly so, yet which demonstrate the principle and the advantages of using the technique.

9.2 FLAMINGO BREEDING

An example from outside of food security concerns sets the scene and leavens the text. The challenge was how to encourage breeding in a *small* population of pink flamingo, which breed only in the wild when in a *large* population. Sydney Zoo had just 11 birds, so installed 20 large mirrors in the compound to create the illusion (to the birds) of a large population. It worked and birds bred. Almaty Zoo in Kazakhstan planned to repeat this from 2015, as it by then had only 20 birds, imported from Cuba in 2008, and had failed hitherto to get them to breed. Identifying the mirror strategy in Sydney was a stroke of genius by one individual. It was an insight and a solution, which linked the birds in captivity with those in the wild state; this solution did not need a gang of endocrinologists to modify the physiology of the birds through injections or feed supplements. The latter *potential* outcome would have resulted from convergent thinking, to modify the bird directly, thence its behavior; lateral (divergent) thinking led to the modification of the bird's environment, thence its behavior. The latter strategy has the additional apparent advantage of being sustainable.[57]

---

[57]The story of flamingo breeding I came across in the in-flight magazine while traveling from Dushanbe, Tajikistan to Astana, Kazakhstan in 2015. I mention this here in this chapter on lateral thinking, to show how valuable it can be to trawl one's environment for anything that might be useful one day, even if one doesn't realize that at the time.

9.3 SWOT ANALYSIS

9.3.1 Strengths

The strengths of a community are often not recognized in its own estimation. This can be gleaned when the community is asked to conduct a SWOT analysis of itself. Individual or group recognition of his/her/its own strengths can serve to encourage that community, and through independent outsider engagement these strengths can be better captured and harnessed.

A certain degree of lateral thinking is required in order to ask the right questions (qualitative and quantitative), and lead the discussion so that the strengths of any individual or community can be laid bare, and exploited to the benefit of that community.[58] Listening and observing fully is a great way to start such an investigation, prior to (or instead of) embarking on a full-blown analysis.

From a SWOT analysis that I once conducted in the Gaza Strip, it was clear that the group concerned did not realize that what it was achieving was truly phenomenal when viewed objectively, in the context of the constraints imposed on it through the Israeli Occupation. Under such tribulations few non-Gazan communities could match the successes that are achieved by Gazans. The degree of cohesion and resilience of the 2 million-strong community in the Gaza Strip is truly impressive; it is especially remarkable because of the diffuse origins from within greater Palestine of the displaced people now living in the Gaza Strip. The Western concept of resilience is underpinned by the Palestinian concept of Sumud (صمود) (Arabic for "steadfastness"), which emphasizes resilience and resistance across Palestine towards the ongoing Occupation, forced-displacement, land-annexation of their territory, and violence inflicted by "settlers" (see Qleibo, 2017, for more detail).

An unrecognized strength in a very different context is related in Box 9.1.

9.3.2 Weaknesses

9.3.2.1 In the Community

Key weaknesses (constraints) in a given community become better-evidenced through a SWOT exercise (however informal), particularly regarding how the community itself can contribute to its advancement, and what the community *cannot* or *will not* do.

---

[58]Quotes ascribed to both Michelangelo and Auguste Rodin relate to the notion that a statue is already present in the marble; all the sculptor has to do is expose it.

BOX 9.1 A Hebridean Unrecognized Strength

A story from a remote Hebridean island off the mainland of Scotland provides an example of a locally unrecognized strength as recorded by Hutchinson (2011). At the time of the anecdote, more than a century ago, the island supported a dispersed community living in turf-clad stone houses, with no electricity or piped water, one in which subsistence cultivation and fishing provided the food source—similar to the rural condition of many a developing country today:

Hebridean horsemanship was justifiably famous. It was not restricted to men and boys. Early in the 1890s a newly arrived schoolteacher was talking to a South Uist parish priest on the pathway to his church door, when along the road came at full gallop a large white horse with a young woman seated sideways on its back. Just as we reached the gate it stopped and she slid easily to the ground almost before it came to a standstill. I noted she had ridden without saddle or bridle. After a rapid conversation in Gaelic, she put her foot on the bar of a gate, sprang lightly to the back of the horse which immediately started off at full gallop on receiving a slap from its rider. It transpired that she had come with some important message from the priest of Bornish, the next parish eight miles away. On my speaking of her remarkable riding my companion exclaimed "Oh, that's nothing", as if it were not worthy of comment.

From my own experience in 2011 in Liberia comes a story of rustic cocoa bean driers. One of the agricultural investment projects I helped formulate for the Liberian government was for the cocoa crop in the north of the country. Due to the two civil wars which swept the country between 1989 and 2003, and the associated human displacement and disruption to the rural economy, smallholder cocoa plantations were in dreadful shape by 2011. Many of the plantation owners had been killed or had not returned from exile to claim their land, the trees consumed by undergrowth or tended by those who had settled on the vacant land who may have had no experience with the crop.

Following fermentation of the beans, their moisture content needs to be reduced from around 60% to 7.5% as quickly as possible so that off-flavors do not develop, as these lower the beans' market value. One of the many ways in which the cocoa value chain could be enhanced in Liberia is for this drying to be done on stacked racks under transparent plastic cover. Compared with the traditional method of bean drying in the open air, using such ventilated solar sheds reduces drying time to about a week, and protects the beans from any late rainy season showers. My Liberian colleague (from the private sector) and I visited one cocoa area to which a solar shed

had been provided as a demonstration. Some 30 farmers had been invited
to meet us. The following is a rough transcript of part of the conversation
which took place:

Farmer: *We want you to show us how to build these.*
Myself: *I don't wish to embarrass you but let me ask you a question. You live in a
house, don't you?*
Farmer: *Yes.*
Myself: *Who built that house? You did, didn't you?*
Farmer: *Yes.*
Myself: *If you can build a house, you can build a shed. It requires only plastic sheet-
ing and nails from the local trading center, together with wood which you can cull
from the forest. I've spent only two minutes inside the shed, during which time I've
understood the structure and could build it myself.*

Both my Liberian colleague and I were of the opinion that the farmers
we were with, and Liberians in general, need to be far more proactive in
helping themselves, rather than relying on outsiders to do something for
them. Yet many are suffering from DPA (Displaced Person's Apathy) and
through their inertia are stuck in a helplessness mindset (Bakis, 1955). Over
my many years in Africa I've not come across a country in which this help-
less condition is so pervasive. Photos 9.1 and 9.2 show how such inertia can
exhibit itself in camps for IDPs too, in this case in Rakhine State, Myanmar.

Photo 9.1 Repairs under way of a broken hand-operated water pump, at a Muslim DP
camp near Sittwe, Rakhine State, Myanmar. Only when the author made persistent en-
quiries as to why the pump had not been repaired did the resident technicians (on the
payroll of an NGO) rouse themselves from slumber and come to repair it—a 10-min task
of replacing the rubber diaphragm (February 14, 2017).

Photo 9.2 At the same DP camp, a public toilet remains broken, for the want of a few bricks or stone to prop up the platform at the front, again exemplifying DPA (Sittwe, Rakhine State, Myanmar) (February 14, 2017).

By contrast with Liberia, Uganda, also caught in violent turmoil lasting 14 years (1972–86), rebounded immediately back to its former entrepreneurial self as soon as the last dictator was ejected. Liberia must aspire to be self-sufficient too, and donor assistance must encourage this resilience rather than going along with a "donations" mindset.

Such a self-help group mentality needs to be encouraged in such fractured societies, in which community trust has been decimated through war, and private initiative destroyed. A direct technical approach on its own may not yield sustainable results. A mixture of availing both a psychological- and business-support system *and* quick results-oriented market-led technical interventions to improve livelihoods and incomes is likely the "dream team" scenario.

9.3.2.2 At Administrative and Management Level

In terms of the traditional hierarchical pyramid, it could equally well be that the weak link in a development action can be above the level of the local community, at local or federal government, say. In Section 4.2.2.4 are some comments on foreign agencies and how blockages can occur at that level of implementation. Let us here look at an example of a Minister in the government blocking a particular project policy—construction of an access road, say. A letter exchange, articles submitted to the press seeking an improved working environment may or may not work, and direct meetings with the Minister concerned could be adopted as the best strategy.

I recall a workshop on nutrition at Ben Gurion University of the Negev at Beer Sheva, Israel, at which renowned Israeli nutrition specialist Professor Ted Tulchinsky and I (both of us members at the time of the Israel-Palestine joint nutrition working group) agreed how important it was to knock on doors to persuade the usually non-technocratic Minister concerned to listen to a strongly argued case regarding food fortification for instance, so that a beneficial change in policy or action results.

At that Beer Sheva workshop, a Palestinian medical doctor lamented how he had become weary of raising important issues with unlistening politicians, with nothing to show for it. Some lateral thinking may bring rewards in such cases of rather non-responsive politicians. It could be that a one-to-one meeting can be arranged with the Minister concerned, in his or her office in the first instance, followed by a meeting at a less formal level, on the golf course, for example, or on the Minister's farm to which he or she travels at weekends. The key ingredient would be to rationalize to the Minister that for him or her to authorize/endorse a proposed course of action would be in his or her own interest (as well as the nation's). The suggestion would not be made directly, of course, but wrapped in such a way that the Minister comes to see it as being in his or her own interests.

Such deviousness was espoused in a British TV sitcom called *Yes Minister* which ran from 1980 to 1984, in which civil service managers were perpetually endeavoring to persuade a Cabinet Minister that a particular course of action would be good for him, when in fact it was the civil servants who would be the prime beneficiaries. The duplicity involved proved to be a strong draw for TV viewers, scoring highly in audience ratings, and it was apparently the favorite television program watched by the then-Prime Minister Margaret Thatcher. A real-life government Minister at the time conceded, under questioning, that the series "did contain an element of truth".

Politicians may mean well, yet not have the knowledge or time to address some key issues, and may need and welcome help. In pursuit of this, I've drafted policy speeches for Ministers and Presidents in several countries, and wrote a weekly column in a national newspaper to inform and sensitize the educated agricultural community.

Yet truly, Ministers and Presidents don't have an easy ride. "Uneasy lies the head that wears a crown", as the King says in Shakespeare's *Henry IV, Part 2*. I have a friend in Afghanistan, a former Minister of excellent repute, who explained to me how he was blocked from achieving the goals he sought as Minister, by a conspiracy of political maneuvering, which "destroyed my faith in my fellow human beings". Indeed, his whole personality had been subdued as a result of the experience.

9.4 POPULATION MANAGEMENT

Global population is set to double between now and 2050. Already the planet's resources seem stretched to breaking point. I cite here a "case study" of the lateral thinking which could help development professionals. In a discussion I had with a high-ranking public sector professional in Gaza, it was clear that the pressure to have children was directly related to the stress under which the Gazan community is living. The greater the stress, the greater the likelihood of human population increase. The same is found in wild mammals too: an expression of the biological imperative for species survival through producing more. Among those working with birth spacing/planned-parenthood initiatives, this correlation needs to be understood and actions leveraged from it—addressing the "circumstance" (Section 6.2.2). As well as direct ways to address family size (Section 6.4 of the earlier book), an indirect way to decrease human population pressure could be approached through measures to decrease stress. These measures may include creating improved livelihood opportunities through vocational training and market development in services and goods, and mental health initiatives to mitigate psychological trauma.

9.5 BRINGING A NEW IDEA TO A COMMUNITY
9.5.1 A Vision for Eco/Agro-Tourism

Eco-tourism is big business in many countries in Africa, for instance, with wildlife parks and lodges, and beautiful coastal and lakeside resort tourism. It gives tourists a wonderful and inspiring holiday, generates local employment, and can promote wildlife protection and biodiversity. In Liberia, Western Africa, this industry is still to develop, with currently only one National Park (Sapo). The vision does not seem to be there, nor therefore the local experience. When I worked in that country I devised an eco-tourism project proposal, which had excellent gross margin analysis figures. I had included photos of rustic thatched guest accommodation, from a lake resort in Uganda which I have patronized over the years. Yet it elicited no positive response from the Minister or his colleagues, who could not believe that foreigners would ever seek out and pay for holidays living in thatched houses which many Africans have spent their lives trying to upgrade! In Liberia, the eco-tourism industry is not dynamic, with most visitors having little choice but to stay in smart modern hotels close to the services to which they are used at home.

The best "lateral thinking" approach may be for a study tour to be arranged for potential investors (Liberians living in the United States particularly) to visit and stay in such beautiful places in Southern and Eastern Africa, to try to whet their appetite. Another aspect that needs to be in place for this idea to flourish is a fine-tuned hospitality culture, which is absent in Liberia at the moment (related to the civil war explained in Sections 5.7 and 8.4). This is something else that can be inculcated during a study tour.

The value of study tours cannot be overestimated, especially in-country or regional. Indeed, that can be the most cost-effective modality, as long-distance travel and accommodation costs are minimized, as are language and cultural barriers. One of the best study tours with which I have been involved (under the Koshi Hills Development Program, Nepal) was bringing coachloads of vegetable farmers from the eastern hills to a valley traversed by the Rajpath road, connecting Hetauda in the plains with Kathmandu. In this valley there is a very profitable enterprise run totally by local farmers all year round, providing cabbages and other vegetables to markets in both the Nepal *terai* and Kathmandu. The visiting farmers saw what was possible in Nepal, under agro-climatic, market and cultural conditions with which they were familiar, and were able to ask questions of the farming community in their own language. Their instructors were their fellow nationals, and the hands-on teaching was done in the field not in smart hotels; thus the whole exercise was culturally appropriate. What these visiting farmers learned from farmers in the central Nepalese hills was immediately implementable in the eastern hills, at the same altitude and agro-ecological zone, with a good market in Biratnagar in the *terai*.

Such a study tour concept can readily be adapted/expanded into having an agro-tourism slant, for foreign tourists and students to learn the techniques of sustainable agriculture from people who engage in it and on which their food security depends, and they can stay in accommodation locally, looked after by local people. This would generate additional employment and revenues of course. Together with UN Habitat staff I formulated such an agro-tourism project proposal for Area C in the Jordan Valley, for a European bilateral donor. The potential students were able to learn not only about the agriculture of the Jordan Valley, but also its rich archeological and religious history, its people and culture, cuisine, and natural flora and fauna. This is surely a win-win means of bringing resilience to the community.

9.5.2 Rapid-Impact Technical Interventions

Just as I mentioned learning from the in-flight magazine in Section 9.2, so I would encourage all technical development workers to share their knowledge at every opportunity, to embrace every serendipitous opportunity to do so. One "small thing" can make a big difference to the prospects of an area or region, some aspect of knowledge culled from another part of the country or another country. I would classify this as lateral thinking, adding value to one's presence in a region through shared knowledge. This is not supply-led indoctrination. It is offering one more item on a menu of options that the given community may consider prioritizing and adopting.

I've mentioned above the intensive vegetable production study tour in Nepal, a "rapid result" suggestion based on my knowledge that a successful market-led vegetable production model was being practiced there, while it was not there in my project area—I drove past it on the short-cut old *Rajpath* looping road up the mountain, rather than the main road that eastern farmers would take in their long-haul bus routes between the east and Kathmandu, so I saw it but they did not.

Another example I have mentioned previously, in Section 8.7, was the benniseed I saw spilling much of its seed in northern Nigeria. There are non-shattering varieties, such as that which is normally grown in a part of Sudan. There is surely a need for these to be tested in north Nigeria, with the prospect of significant increase in the grain which farmers can actually harvest.

I mention biofortification at every opportunity (Chapter 5 of the earlier book), as for me this is a sustainable means of improving the nutrition of millions. I am neither a professional breeder nor biochemist, so cannot run such a program. Yet I can raise awareness of its potential, citing initiatives ongoing in northern Nigeria, for instance, and that four scientists shared the World Food Prize in 2016 for their work in breeding, and sensitizing the general public about, biofortified crops. Through the combined efforts of the four Laureates, over 10 million people are now positively impacted by biofortified crops, with a potential several hundred million more having their nutrition and health enhanced in coming decades. Both biofortification and non-shattering benniseed are nutrition-sensitive interventions which promote nutrition security, and in turn better health and school and work performance, all contributing to enhanced resilience against food and nutrition insecurity.

Other rapid-impact examples in my experience include showing a farmer in Borno State, Nigeria how the roots of his stunted sorghum had been parasitized by the roots of the *Striga* weed. Over years of suffering yield loss due to *Striga* he had never looked closely at the way the two root systems fused. I invited him to dig up and examine closely the roots of any stunted sorghum plant he chose. He was flabbergasted at what he saw.

Another example—in 1985 while I was formulating an irrigation project in Ethiopia, I was passing Lake Awasa agricultural research station. Rather than continuing my journey, I decided to spend a few minutes with the staff in case I knew something which they could find of use, and indeed learn something new myself. One of the scientists showed me his plots of soybeans, which were very yellow and stunted. I had experienced just the same symptoms myself with soy I had planted in Libya a few years previously. I invited him to pluck any of the plants and snap open the stem. He did and was shocked to find it hollow and brown, which is what I had expected. I relayed to him my experience in Libya, and that the symptom was most likely the result of infestation by the insect *Melanagromyza*, which had been identified for me by my entomological colleague at the agricultural field research station at Tajoora, near Tripoli. The adult deposits her egg into the stem of the young plant. On hatching, the larva then burrows through the soft tissue so that sugar translocation from leaves to roots is prevented. The Ethiopian scientist was aghast at the simple explanation which he had not come across before. I left him thinking of how, once he had verified the cause of the symptoms, he could control this pest biochemically, or avoid it through a soy variety selection program.

Rapid technical-impact interventions can also come about through a physical input, such as a pedestrian rope bridge to span a gulley or river, or a pre-cast concrete ring culvert to render a road all-weather, both common in the low hills of Nepal. The Gambian farmer shown in Photo 9.3 is in need of a quick-fix solution too, to weld a broken blade on his diesel-driven rice paddy cultivator. My field party met him in Sukuta, Western Division. The gun he is carrying is to scare off hippos which emerge from The Gambia River and damage his rice crops.

Photo 9.3 Gambian farmer in the tidal-irrigated rice scheme at Sukuta en route to the village blacksmith (June 6, 2017).

An example of a rapid-impact intervention from Sudan is the outboard engines and ice boxes provided under the EU-funded Sudan Food Security Program, which transformed the fortunes of the artisanal fishing communities on the Red Sea near Port Sudan. Until then, their small inshore boats suffered a severe market disadvantage at the hands of the larger foreign-owned boats fishing further out to sea and landing their catch in Port Sudan (Photo 9.4).

9.5.3 Differing Perspectives

In Section 6.4, I included my poem published in the *Palestine-Israel Journal*, seeking to convey contrasting perspectives of both Palestinian and Israeli Jew to collateral damage caused to both sides in the conflict. The latter represents an enormous burden on Palestinians, and also Israelis, and their supporters who fund them. Were that money to be spent on development initiatives rather than building the Separation Wall, military hardware and humanitarian aid, the number of food-insecure in the world would decrease significantly—a peace dividend.

Photo 9.4 A gathering of 16 members of the Suakin Fishermen's Association, Red Sea State, just south of Port Sudan. The group is briefing a member of the sub-Program team from the implementing NGO, SOS Sahel (April 23, 2016).

Two other examples from the same part of the world—I once offered to a very liberal Israeli devoted to promoting better understanding of "The Other" in Israel my perspective on a "problem" as seen by traditional-minded Jews in Israel. This relates to their aversion to the increasing tendency of Jews in the United States to marry outside of their religion; this is perceived to be a regrettable trend, against which lobby groups have railed. I opined to my Israeli friend that I did not see it that way at all, and to the contrary, had I been a Jew myself I would have been flattered at this trend. First, it shows that non-Jews in the United States can love "us". And second, far from representing a dilution of Jewish culture, it provides Jews with a means of outreach into non-Jewish society whereby "my" Jewish beliefs and norms may be communicated to a wider audience, and surely this would be a good thing? My friend was pleasantly shocked at my perspective, saying he had never thought about it like that before.

Debate, whereby ideas are shared for consideration by others, is surely a good thing too, with, hopefully, good ideas winning the day over bad ideas. Fostering greater resilience against food and nutrition insecurity is a case in point. It may take a while for "good ideas" to be accepted; that is part of the course. After spear-heading the formulation of the national Food Security Strategy in Palestine in 2005, I considered that such a strategy was needed in Israel too. Through my involvement at NGO level with the peace process in the region, I had become friends with a senior member of a liberal political party in Israel. I suggested to him that the time was now ripe for a national Food Security Strategy, for neighboring Palestine had already devised one, and an action plan to implement it. He replied that the idea was a great one; however, it needed cross-party political support in Israel, and as the country was just approaching a general election (February 2009), cross-party support on anything was highly unlikely! I came to understand some years later that formulation of Israel's National Food Security Strategy had started.

9.6 THE VALUE OF CORN COBS IN A PARCHED FIELD

On the front cover of the current author's earlier book is a picture of a desiccated corn cob, with a low grain set. The plant is surrounded by others which do not carry a cob at all. It is to be hoped that despite the farmer's great need for food following that harvest season failure, he or she was able to keep this cob to use as seed during the following season. That plant had managed to use the soil moisture to a better extent than had neighboring plants, due to a few days' earlier maturity maybe (drought avoidance) or an heritable factor which had conferred physiological tolerance to drought. If, because of drought, few other plants had shed their pollen during tasseling in that field, thereby reducing the prospect of cross-pollination, there is a good chance that those seeds would give rise to drought-tolerant plants, with a competitive advantage.

Even if the farmer and his or her family wishes or has to eat the grains on the cob, just maybe an observant Agricultural Development Officer, for example, could be passing before the cob is harvested. The farmer could be offered a large cob or other inducement in a barter arrangement and the ADO might grow the seeds in a demonstration/multiplication plot the following season. In the earlier book I cited a field of exceptional wheat I came across in the hills of eastern Nepal, and I offered the farmer an equal quantity of rice if he would avail me of his harvest, so that I would be able to test his variety as a test entry in PVS trials the following year (Ashley and Khatiwada, 1992). We did a deal on that basis. Promoting that type of alertness among agricultural staff would be worth pursuing.

9.7 POTATO PROMOTION IN FRANCE

So-called "European potatoes" were initially brought into Europe from South America, at the beginning of the 16th century, following Spanish colonial conquests there. Their subsequent spread in France is a story worth telling in this chapter, as it involves some serious lateral thinking by its main champion, Monsieur Parmentier.

Prussia seems to have adopted the potato as a human food early on, with mandatory pressure to cultivate the crop from Prussian King Frederick II applied to the peasant class, to counter a blockade of grain imports to Prussia by France. Otherwise, apart from Spain and Ireland, European countries were initially reluctant to adopt the crop, and France in particular seems to have been among the most reluctant to do so. The plant is related to the poisonous weed deadly nightshade after all, and humans eating the above-ground parts of potatoes seem to have been stricken. The idea of humans eating the part of the crop that was underground was alien too, as these potato tubers were also assumed to be poisonous, with some saying that they caused leprosy. The French government was so concerned about the health effects that the growing and human consumption of potatoes was outlawed by the French Parliament in 1748. The potato was fed to pigs in Spain and Ireland though.

The crop's fortunes rose, however, as a result of Monsieur Antoine-Augustin Parmentier, a French army pharmacist and nutritionist, who had the misfortune of being taken prisoner-of-war by the Prussians in 1771 during the Seven Years' War. During his incarceration in Prussia he was fed with potato tubers, yet noticed no ill effect. On his release he returned to France and spread the word that the potato was a new crop which could be eaten by the French, and used to better nourish patients with dysentery. He also campaigned for potatoes as a substitute for grain when those crops failed, suggesting potatoes could be turned into flour too. His advice was taken, the potato thereafter serving as a vital alternative staple in times of grain harvest failure, thereby preventing famine and adding further credibility to the potato.[59] As a result of his awareness campaign and lobbying, the ban on the potato in France was lifted in 1772, with the respected Paris Faculty of Medicine pronouncing the crop as edible the same year. Thereafter, the crop's fortunes were transformed, and within a couple of decades the crop was widely grown in France, including in the royal gardens at the Tuileries Palace in Paris, where some ornamental areas were replaced by potato plots.

---

[59]It was my history buff mother who explained to me when I was 8 years old that during the (subsequent) French Revolution (1789–99), the poor people who were able to find and eat the potato peelings discarded by the rich were better able to survive the food and nutritional insecurity which widely accompanied the political and social upheaval of the time.

Parmentier implemented a wide range of awareness-promoting measures, and proved a marketing supremo. He sought support from the royal family for his cause, presenting King Louis XVI and Marie-Antoinette with bouquets of potato blossoms, leading to Marie-Antoinette wearing the flowers on her hat. The King commended him for having "found food for the poor". Parmentier also hosted potato-based dinner parties for luminaries who could broadcast the value of the potato in their turn, such as Benjamin Franklin (one of the Founding Fathers of the United States) and Antoine Lavoisier (the "father" of modern chemistry).

As a master stroke of pure genius, Parmentier planted and publicized his own field of potatoes at Sablons (the land a gift from the king), and as the crop ripened, he mounted an armed guard around the potato plot, thereby conferring a value on the crop. He instructed those guards, however, that they were entitled to accept bribes from anyone who wished to steal some tubers during the daytime! Moreover, at night, the guards were intentionally withdrawn, returning only in the morning. Before long, the mature potatoes had all been dug up by nocturnal thieves who had internalized Parmentier's advocacy message, and then planted the tubers in their own plots of land (to eat and/or for multiplication and sale presumably, each comprising a sound personal food security strategy).

9.8 NEPAL EARTHQUAKE IN 1998

On August 21, 1988, there was an earthquake in Nepal of magnitude 6.8, in which 722 people in the east of the country died. One of the worst-affected areas was Koshi Zone in the high hills and low slopes of the Himalayas, where I was then working, and our project vehicles played a prominent role in collecting the injured who had been brought from the interior to the roadside. King Birendra visited the area, once his palace officials thought it safe for him to do so. He needed transport once he alighted from his helicopter at the grassed parade area in Dhankuta (2,547 ft, 776 m), and my personal project vehicle was volunteered for this purpose.

My project was funded by the British government, and our fleet of white Land Rovers each carried a small British Union Jack flag on the driver's door. This was of diplomatic concern to the palace officials. They decreed that the king could not ride in a vehicle which had another country's flag on it, so it was peeled off to appease sensibilities.

However, Mr. Gurung, our project's chief driver, thought laterally. He was, like all of the project fleet drivers, of ex-British Gurkha regiment pedigree, and felt he was as much British as he was Nepalese. With admirable adroitness, he procured an enormous stick-on Union Jack, presumably from his mates in the British Gurkha garrison in Dharan (1,148 ft, 349 m), in the *terai* at the foot of the eastern mountain range, three-quarters of an hour's drive from Dhankuta. Mr. Gurung attached this flag to the underside of the roof in the vehicle in which the King was to ride, so it was highly visible to those inside the vehicle, though not to anyone outside.

The King was duly collected from his helicopter and driven to inspect damage incurred in the earthquake, as could be easily accessed on the permanent road which had been constructed by the project from Dharan through Dhankuta to Khandbari at 2,844 ft (867 m), most of this by then tarmacked. His Majesty could not but be aware of this flag every minute he was in the vehicle, and at some point on the journey asked driver Gurung, "Do you like working with the British?". "Yes Sir", was the reply. The late King Birendra was highly intelligent, and would have appreciated the sleight of hand that the "British", who had hosted him during his schooldays at Eton, had devised to place the flag above his head. He would also have seen first-hand the assistance that the British government had brought to the area, over an extended period, not just during the earthquake.[60]

The King's understanding of this was important, as the Nepal government's political line at the time had been to downplay the damage wrought by the earthquake, maintaining that the country did not need foreign assistance to redress that damage. A British RAF Hercules transport plane carrying aid supplies for the victims was for a long time refused permission to deliver those supplies at the regional airport in Biratnagar. It was eventually authorized to set off from India, land and discharge its cargo. I happened to be at the airport when the empty plane took off, and witnessed a spectacular display. Having become airborne, and surely without permission from the control tower, the pilot turned sharply and swooped very low and noisily over the airport building, dipping his wings. I felt that in so doing, the pilot was making a rude gesture at the Nepalese government for its delayed authorization for the plane to bring in the supplies. Another example of lateral thinking perhaps, from a pilot, not a driver, this time.

---

[60]Immediately following the earthquake, I had cabled the Dharan British Ghurkha garrison hospital from Dhankuta and asked if it could accept the civilian casualties we were collecting in our vehicles. The hospital's laconic reply was, movingly, "Any time, any number". The government hospital in Dhankuta was not equipped to handle serious cases, whereas the army hospital at Dharan was.

Maybe driver Gurung's lateral thinking played a part in ensuring that the Koshi Hills Development Program received national buy-in at all levels, and recognition that its development emphasis, together with its humanitarian help in 1998, were expressions of the friendship and historical links between the two countries. Many of the British (and Indian) Gurkha soldiers over the years have been recruited from Nepal's Koshi Zone; Mr. Gurung and others together with their extended families were among the beneficiaries of that and the Koshi Hills Program.

REFERENCES

Ashley, J.M., Khatiwada, B.P., 1992. Local wheat makes the grade in Nepal. Appropriate Technol. 18 (4), 30–32.
Bakis, E., 1955. DP apathy. In: Murphy, H.B.M. (Ed.), Flight and Resettlement. UNESCO, Paris, pp. 84–85.
Hutchinson, R., 2011. The Silent Weaver: The Extraordinary Life and Work of Angus MacPhee. Birlinn, Edinburgh, pp. 50–51.
Qleibo, E., 2017. Vivre à Gaza: pour une sociologie de la résilience (Doctor of Sociology thesis). University of Burgundy-Franche-Comté, France. November.

CHAPTER 10

The Role of Champions

Contents

10.1	Champions at Village and Public Sector Levels	167
10.2	Champions From the Commercial Private Sector	168
10.3	Champions Who Contest the Commercial Private Sector	170
10.4	Champions From International Organizations, Sport and Entertainment	171
References		172

It is important to identify local champions to facilitate development of resilience in a given location; this is best done on the basis of their actions, not words. These champions will be respected by the community because of their proven knowledge, skills and wisdom, and their ability to make good decisions over and over again. All this gives them credibility and influence. Their voice would be heeded by those in administrative and political authority. The importance of these champions to the success of an intervention cannot be overestimated. Interventions that are planned can be tested against these champions, to assess whether they would fly. Indeed, the champion may likely be able to suggest interventions that would be useful, which an outside project team would not think of.

10.1 CHAMPIONS AT VILLAGE AND PUBLIC SECTOR LEVELS

Champions are good at creating facts on the ground ahead of public opinion buy-in. Examples given in the earlier book included Mr. Jawali, who demonstrated the huge benefits of animal draft cultivation in the hills of western Nepal, to replace manual work, hitherto largely done by women. Such champions are common at village level, for instance, local traditional leaders, who are often held in higher regard than elected or appointed officials within the national administration. Often they keep a low profile, but they are there for consultation, happy to share their wisdom and know-how.

In 1999 at Makurdi in Nigeria, on a project funded by the Petroleum Trust Fund (PTF), I called a workshop to hear the views of people in

Benue State on how the government agricultural extension service could be improved. The star attraction at that workshop was a Nigerian man who had worked with the colonial administration agricultural department (see Section 8.9).

In the public sector, the Minister of Women's Affairs in Afghanistan (Husn Banu Ghazanfar) impressed me in 2010 with her determination. I was similarly impressed by Vabah K. Gayflor, the Minister for Gender and Development in Liberia in 2011; I recall in particular her anger at the head of one of the donor missions in Monrovia who showed peremptory disinterest in attending any function which her Ministry organized to spread gender awareness.[61]

10.2 CHAMPIONS FROM THE COMMERCIAL PRIVATE SECTOR

Champions are also to be found in the commercial private sector—for instance, **Dr. D. Warren Harrison**. Following earlier work in Ghana, Dr. Harrison, an African-American Seventh-day Adventist missionary doctor, came to work in Uganda in 1964. He was originally hired by the Uganda government as Head of Nutrition and Health Education, with an emphasis on local commercial production of high-protein nutritious foods. He set up a soy steam extrusion factory in Kampala to deactivate the antitrypsin factor, the resulting product being used to make a range of secondary products, especially targeting the poor. The SME was called Africa Basic Foods, initially run as a non-profit corporation.

This admirable initiative led to the uptake of proteinaceous soy/maize mix flour porridge in urban households in the country, suitable for weaning and children in particular. It provided a market for locally-grown soybeans, and this farming enterprise was supported by the soy improvement program at Makerere University and the government research station at Kawanda (both started on the initiative of Dr. Harrison). Harrison and his Ugandan counterparts provided the value-adding link between farmers and consumers. Lessons were learned from the exercise, including the need for better buy-in from institutions like schools which could provide a market, and that the market in Kampala for the new crop soy was constrained by the reluctance of the majority Baganda there to eat maize compared with their normal staple, *matooke* plantain.

[61]I was delighted to mention it in my report to that donor at the time, which I hope discomforted him.

Dr. Harrison funded much of his work in Uganda through his part-time medical practice in Kampala, with the support of UNICEF, the NGO CARE and well-wishers. The enterprise suffered during Idi Amin's dictatorship (1971–79), and in 1974 Dr. Harrison moved to neighboring Kenya, again working with soy.[62] I met him in Kampala just before he left for Kenya, not realizing at the time what an important influence he had had in Uganda. His threefold approach of encouraging soybean production, starting viable businesses to employ people making soyfoods, and providing grass-roots health education proved a practical way of helping local people on many levels. There is no doubt of his status as a champion, seeking neither fame nor fortune, but rather simply bringing better health and income generation capacity to the people.[63]

Another example of a champion from the commercial private sector is **Mazen Sinokrot**, a politically-independent and non-partisan business leader in Palestine. He is the founder of Sinokrot Global Group (SGG), the leading family-owned business in Palestine with regard to investment size and number of employees. SGG focuses its investments in food and agro-industries, tourism and hospitality, trade and information technology. I first met Sinokrot in 2003, identifying him as a key stakeholder from the private sector who should participate in the formulation of the Palestine National Food Security Strategy. His group of companies includes one which tins sardines. I asked him why he doesn't use Palestinian olive oil in these, to support local olive growers. "I can't afford to", he said, its local price being higher than imported vegetable oil. That was a window on the uncompetitive state of the Palestinian olive oil subsector, which remains at subsistence level to this day, though it is receiving attention downstream of the formulation of the national olive subsector strategy.

Sinokrot led the national effort to establish a number of private sector institutions including the Federation of Industries, which he chaired for a full term. He was founder and first chairman of the Palestine Trade Center (Paltrade), and serves on the boards of a number of national investment and university institutions, and foundations that promote youth development, public participation and social responsibility. He has been chairman of the boards of the Palestine Standards Institute, the Palestine Investment and Promotion Agency and the Palestine Industrial Zones and Free Zones

---

[62]By 1983, Africa Basic Foods had returned to good health. In 1997 it was renamed as East Africa Basic Foods.

[63]http://www.soyinfocenter.com/HSS/harrison_and_africa_basic_foods.php.

Authority. Sinokrot is a major promoter of public-private partnerships, facilitating formulation of laws that encourage a competitive economy.

I identify Sinokrot as a champion for having created so much local employment, and by sheer determination managing to both expand his business empire and promote national economic interests in tandem, despite the impediments imposed by the Occupation. I met him again later when, as well as still heading his business interests, he was Minister of the Economy (2005–06). In the presence of then-Prime Minister Salam Fayyad, he railed against the Occupation—"Are we Palestinians meant to live underground?"—to make way for the Israeli-only roads to service the Settlements in West Bank. Such forthright statements make him popular among Palestinians as a figurehead who speaks his mind.

10.3 CHAMPIONS WHO CONTEST THE COMMERCIAL PRIVATE SECTOR

The broadest definition of *food activism*, says Joan Gross (2017), is bringing food practices into consciousness with a goal of changing them. She suggests that food activism becomes more interesting to anthropologists as it moves beyond the scale of the individual (one's own food habits). The target group can expand to include those of one's family, college campus, community, nation or global family. Food activists may work to improve the health of the environment through its role in food production on the one hand, and the health of people as food consumers on the other. Such actions often involve analyzing the conditioning psychologies, economics, policies and politics. Many food activists working at the larger scale today started off by consciously changing the way they ate; such change had its roots in the political ferment on the streets of the United States in the 1960s (Belasco, 2007).

Ethnography, with its cornerstone of participant observation, opines Gross, is a cogent means of studying how people can organize themselves to make changes related to production or consumption in the world, referring to the presentation of 14 cases of food activism (local, national and international) in a book edited by Counihan and Siniscalchi (2014). This volume demonstrates that food activism can have wide-reaching effects, affecting power structures and injustices, such as land grabbing, increasing taxes and privatization of land, and domination and oppression by governments or local elites, all conditioning resilience against local food insecurity (see Sections 6.1 and 11.1).

The range of food activists' immediate goals is enormous, including advocating school meals in marginalized communities, fostering "ethical consumption" concerns involving Fairtrade/corporate responsibility and workers' rights, animal welfare and food sovereignty concerns, or promoting adoption of "slow-food" or organic agriculture as alternatives to mainstream Western agricultural production and consumption.[64] The "alternative" champions are often adversaries of multinational companies and their dominance of marketing and distribution networks. These food activist champions can vary in scale from small and localized to huge and international. The larger champions have organized themselves well around their chosen theme, and in turn seek to organize others to create stronger bonding between producers and consumers, in the interests of nurturing a more sustainable and equitable global food system. The disaggregated rationales of such activists include better health for consumers, a more sustainable physical environment which nurtures farming and biodiversity, improved social and ownership links within the community, and better remuneration for smallholders who produce most of the world's food. The latter brand of activist likely addresses direct sales by producer to consumer, with value addition—cleaning, processing, packaging—being undertaken by the producer, thereby providing both better margins for the producer and more affordable prices for the buyer.

10.4 CHAMPIONS FROM INTERNATIONAL ORGANIZATIONS, SPORT AND ENTERTAINMENT

Champions in this final category include **Harvest Plus** (part of the CGIAR Research Program on Agriculture for Nutrition and Health) and the **Bill and Melinda Gates Foundation** being champions of biofortification initiatives, such as the biofortified high provitamin A cassava in northern Nigeria, a prime example of building community resilience against nutritional insecurity. The biofortification initiative in Nigeria is supported by local popstars, who have made promotional videos to increase the concept's appeal.

The potential for such stars has been harnessed by the UN system, with well-known international entertainers being drafted in as celebrity Goodwill Ambassadors, for organizations such as UNICEF and the WFP. In 1999, Ethiopia's **Haile Gebrselassie**, regarded as the greatest distance runner ever, was named a goodwill ambassador for UNDP in Ethiopia. In 2012

---

[64]http://www.slowfood.com; http://www.viacampesina.org/en (International Peasant's Movement) (accessed August 15, 2017).

in London he was one of the high-profile attendees at the Olympic hunger summit, an international gathering on August 12, the closing day of the Summer Olympics in Britain, a meeting chaired by the then British Prime Minister.

The appeal of Gebrselassie I witnessed first-hand once. I was on a plane with him, and saw the total adoration for him at the airport in Addis Ababa when he alighted. He was immediately recognized by every Ethiopian and treated as an icon, though he was dressed casually, wearing trainers and acting humbly, prepared to converse with anyone. Everyone felt it an honor to be in his presence, a world-class figure who had excelled and brought recognition of his country for the "right" reasons, as a counterbalance to its negative portrayal in the media as a country associated only with drought and famine. His support for any development initiative has to be helpful.

I would like to mention the British novelist J.K. Rowling here too. I talked earlier (Section 4.4) of "success breeding success", but Joanne is a supreme example of "failure breeding success". Her early years of poverty and struggle for recognition, with many publishers turning her down, consolidated her resolve to continue doing what she believed in, until, with a serendipitous turn of fortune, she started her steep ascent to stardom. Her triumph over early hardship is captured in the Commencement speech for new graduates that she gave at Harvard University in 2008, in which she talks of this resolve being strengthened by early failure. That resolve and determination is a lesson to be well-learned by development workers. Expect setbacks, expect disappointment, expect lack of recognition—but "all in the end will be harvest".

In that Commencement speech she also spoke of the importance of imagination, which proved invaluable to her in understanding and empathizing with the experiences of others, despite not having shared those experiences herself. She gives the example of what she learned while working with Amnesty International. I commend to all who endeavor to help those who are marginalized and food-insecure that you read that speech, for it may give you the key to enter and empathize with their world. With that key you will likely be better able to change their reality than had you not exercised greater imagination!

REFERENCES

Belasco, W., 2007. Appetite for Change: How the Counterculture Took on the Food Industry. Cornell University Press, Ithaca, NY. 327 pp.

Counihan, C., Siniscalchi, V. (Eds.), 2014. Food Activism: Agency, Democracy and Economy. Bloomsbury, New York.

Gross, J., 2017. Food activism. In: Chrzan, J., Brett, J. (Eds.), Food Health: Nutrition, Technology and Public Health. Berghahn Books, New York, pp. 106–117 (Chapter 6).

CHAPTER 11

Case Studies

Contents

11.1	Case Study 1. The Need for More Resilient Food Systems	173
	11.1.1 Introduction	173
	11.1.2 The Challenge of Maintaining Soil Fertility in Perpetuity	177
	11.1.3 The Challenge of Maintaining Social Cohesion and Socio-Economic Integrity	186
11.2	Case Study 2. Resilience to Food and Nutrition Security Among the Inuit	190
	11.2.1 Introduction	190
	11.2.2 Food Security Strategies	192
	11.2.3 Food and Nutritional Security of the Traditional Lifestyle Compared With That in the Modern Settlement	197
	11.2.4 Overarching Inuit Cultural Factors Affecting Resilience	200
11.3	Case Study 3. Human Capital as a Resilience Strategy in the Pamirs	202
	11.3.1 Political and Economic Context	202
	11.3.2 Educated Cadre of First-Generation Farmers	203
	11.3.3 Willingness to Engage in Modern Development Initiatives	204
	11.3.4 Country-Wide Context	205
References		207

11.1 CASE STUDY 1. THE NEED FOR MORE RESILIENT FOOD SYSTEMS

11.1.1 Introduction

Food insecurity is rife, and the reasons for this have been discussed in Chapter 3 of the earlier book by the current author. This section is a continuation of Section 6.1, and considers the fundamental concept of sustainability of global "food systems", which seems to be breaking down; the 2008 food price crisis brought that into focus, thereby leading to a threat to resilience globally.

This case study examines two of the main reasons for the impending structural collapse. Let it be said to start with that neither of the two challenges advanced below is insoluble, yet time to redress both is limited.

The *first* challenge is to ensure that the soil on which our crops grow, and which provides food for the livestock in our food chains, can go on supporting current levels of production at the needful levels. Is the conventional food production system capable of providing food for us in perpetuity, not just for the current 7 billion people, but for the far higher numbers of people the planet is required to feed in future? Or are "alternative" production systems required which would be more resilient—if there are any?

The *second* challenge is to ensure that the food that is produced can be accessed by us all, not just by the wealthy who can afford it, should its availability decline. Are alternative ways of distributing and marketing food needed to underwrite equitability and the "right to food?". Not least of the concerns is the potential threat to global food security conditioned by "choke points" on the waterways of the world (Bailey and Wellesley, 2017). I endeavored to address that threat represented by the Strait of Hormuz while helping to develop the food security strategy for Kuwait, yet that was "only" a regional choke point. What if there were an "event" at the Panama Canal, caused by natural calamity, conflict or institutional breakdown, causing it to be out of action for a long period? That would likely create a huge food price crisis, as the canal controls much of the flow of grain between Western and Asian markets.

These two challenges are interrelated. At every step and turn as the way forward is charted, decisions are needed, at the macro and micro-level, the sum total of these determining global and individual resilience against food insecurity, which this book seeks to highlight.

On the *first score of sustainable "soil fertility"*, which is the fundamental component of "soil productivity", the urgent need to look for alternatives is predicated on (a) the modern high-input high-output industrial-level farming system, which is prevalent in the developed world, this having already led to environmental disorder and social dislocation, and (b) the fact that the context which enabled the mainstream approach to flourish no longer exists—the population has grown, and resources for agricultural/livestock production are more limited and/or more expensive, through scarcity and despoliation.

The mainstream method has successfully enabled food production to exceed the incremental need for food due to population growth over recent decades, something that the earlier book by the current author detailed in its Section 5.3.1. There are those who believe that advances in genetic engineering and developing crop varieties tolerant of salinity or drought will continue to provide similar production and productivity increases in the immediate future (Fedoroff et al., 2010). Yet the "writing is on the wall" that an intensive agricultural system has a finite life, as displayed for instance by the dustbowl phenomenon in Nebraska, United States in the 1940s, and the UNEP report of 2007 which pointed to catastrophic long-term consequences for Sudan if intensive agricultural attrition continues on the fundamental resource of soil.

On the *second score of food technology, distribution, processing and marketing,* the liberalization of international trade and global competition has led to increasing vertical integration of the food chain, which has been blamed by many for causing various kinds of damage to the environment, failure to provide wholesome and nutritious and safe food, inability to supply food for low-income people, as well as for various other social ills (Allen, 1999; Scialabba and Hattam, 2002). As Sen indicated in his essay of 1981, as a result of competition for control of the food market, one group can be disadvantaged because of another's success (Sen, 1981). The Great Bengal famine of 1943–44 which he was addressing did not hurt everybody.

Another example of this unequal opportunity and outcome is the Green Revolution from the 1940s, during which richer farmers did well but poorer ones did worse, as they could not afford the fertilizer needed for the introduced high-yielding varieties to prosper (Box 6.1 above). The Green Revolution did not well address social equity, gender justice or chronic unemployment, which went "missing in action" in deference to the biological imperative of fertilizer-hungry high-yielding varieties. The social context imperative has also fallen by the wayside in the 2016 Global Agricultural Productivity (GAP) Report from the Global Harvest Initiative (representing the views of huge technological companies), subtitled "Sustainability in an uncertain season" in which only the need to produce more food is the message, rather than equity aspects of its production and distribution.

Cleveland (2014) addresses the practices and consequences of our current global food system, in an endeavor to find ways to transform it to one that is truly more sustainable. His treatise provides critical thinking and suggested effective action for the future of our global food system, based on an understanding of the system's biological roots and sociocultural values. It involves an analysis of the assumptions underlying the various perspectives on challenges related to food and agriculture around the world, and a discussion of alternative solutions. Cleveland points out that the mainstream agrifood system monopolizes the bulk of research and development resources, so disadvantaging those espousing and devising alternative scenarios. He details some alternative contesting foodways, and highlights the perspectives of advocates of alternative agrifood systems who argue that demand can be lowered through better diets and reduction of waste. The future resolution of the contest will be influenced by decisions at government, smallholder farm or landowner level, and consumers-cum-general public, just as national- and subnational-level decisions on accepting Genetically Modified Organisms (GMOs), such as Miracle Rice (IR 8) have already been influenced.

The current "conventional" (mainstream) and "alternative" perspectives can sound superficially similar, yet they differ fundamentally in terms of their assumptions, problem definitions and solutions.[65] Alternative more socially-sustainable food systems have been suggested as a counter to the shortcomings of the industrialized and increasingly-global food system (Sumelius and Vesala, 2005).

Over the last couple of decades, there has been considerable interest in developed countries in what have been variously labeled as "alternative food strategies", "alternative food supply chains", "alternative consumption practices", "alternative food networks" and the "alternative food economy" (Morris and Kirwan, 2012). These labels are used to describe a number of diverse initiatives and developments that have recently risen to prominence in the agrifood system. Examples include organic and other forms of eco-logical agriculture; direct marketing such as farm shops, farmer's markets and box schemes and fair-trade produce from locally unique and distinctive places of production; and, consumer lifestyles which condition foodways from the demand side, such as vegetarianism. The term "alternative" sets them in opposition to the current mainstream global agrifood system, as the "alternative" protagonists offer more sustainable, environmentally-friendly and more equitable solutions as they see it. In the alternatives proposed, the relationships between producer, marketer and consumer are closer than in the conventional model.

There has been a political agenda in parallel with these social move-ments, to wrest control from corporate agribusiness and/or add value to the rural economy, into which profits are reinvested. Such a reversal has been termed "economies of scope" compared with the dominant post-Second World War "economies of scale" (Marsden et al., 2002). There is a concern, however, that some of these alternatives are unaffordable for those with lim-ited disposable income and/or cultural capital (Goodman, 2004).

---

[65]https://en.wikipedia.org/wiki/Food_systems (accessed September 5, 2017) comprises a succinct and balanced introduction to the concepts of conventional and alternative food systems.

11.1.2 The Challenge of Maintaining Soil Fertility in Perpetuity
11.1.2.1 The Conventional Modern Agriculture Scenario

As indicated above, a soil health crisis is in the making in the developed world, rooted in industrial level farming methods; this also applies to those parts of the developing world where these methods pertain (such as in the plains of Sudan). Soil is the dynamic natural body on much of the land surface of the planet in which plants grow, providing there is sufficient moisture; it comprises mineral and organic materials of varying amounts, and life of various forms. However, global soils are in retreat, in terms of their capacity to continue providing us with the bounty that the world needs, especially in the context of climate change on the one hand, and a population that is predicted to grow from the current 7.3 million to 9.7 million by 2050 on the other.

Though the current book focuses on the *developing world*, soil conditions in the *developed world* need to be scrutinized—if agricultural production and productivity of staple crops decline there, then grain supply will need sourcing to a greater degree than currently from the developing world. The price per unit of grain which the developed world would be prepared and able to pay in that scenario would be more than that affordable by people in developing countries. Considering normal supply and demand economics, a shortfall in food in developed countries would therefore likely decrease that available in developing countries, the latter also experiencing soil deterioration, and a greater rate of population growth than in developed countries. So it is in everyone's interests that soil fertility in developed countries be maintained.

Lymbery (2017) lists the chronology of farming's industrial revolution over much of the last century. The capture of atmospheric nitrogen into ammonia, then used to make inorganic fertilizer, was followed by US government support to American farmers, downstream of the 1930s Great Depression. A massive increase in corn (maize) production resulted. Subsidized cereals became so cheap and readily available that they were used as livestock feed, with erstwhile pastures being turned into yet more cereal fields for yet more feed for zero-grazed livestock. After the Second World War, US-style industrial farming methods reached Europe via the Marshall Plan, which itself was designed to lever Europe out of its ruined condition. The US Aid package enabled Europeans to purchase food from "Uncle Sam", and thereafter to buy the physical inputs for its own version of intensive farming. The result was monocultures, drenched with poisonous agrochemicals, which created an environment in which biodiversity in the fields perished, the crops subject to pest attack not kept under control by their natural enemies. Inorganic fertilizers replaced the erstwhile tried-and-tested ways of maintaining soil fertility. Competition for land ensued, between crops for human consumption and crops to be fed to livestock. At the global level today, one-third of cereal harvests, and almost all the soy, is destined for livestock feed. The concept of there being a human food shortage is a result of so much human food being devoted to feeding livestock, with a massive loss of energy between those two trophic levels in the food chain; there is more than sufficient grain produced in the world to feed the world population, were so much not devoted to livestock feed (and biofuel).

In an article entitled "Ploughing on regardless", published in the *Guardian* newspaper on March 25, 2015, and reproduced on his personal website, the eloquent and ubiquitous political and environmental activist George Monbiot addresses soil loss, considering almost all other issues to be superficial by comparison. He laments bad policy decisions by both the UK government and the UK National Farmers' Union, failing to address adequately the attrition on the UK's soil, leading to its pauperization.[66] Monbiot opines that globalization ensures that this disaster is reproduced everywhere. In its early stages, globalization enhanced resilience, with people no longer dependent on the vagaries of local production. Yet as it proceeded, it spread the same destructive processes to all corners of the Earth, undermining resilience, and threatening to bring down food systems everywhere.

There are many publications that draw attention to the deteriorating condition of the world's soil. Many are written by soil scientists, or syntheses of their evidence-based accounts and predictions on the downstream threat to food supply unless alternative means of food production are urgently enacted. Six of these publications are cited below to show that this is a real threat already upon us.[67] Business-as-usual on the farm is not a rational option.

(A) FAO

If current rates of degradation continue, all of the world's top soil could be gone within 60 years, Maria-Helena Semedo, Deputy Director General of Natural Resources in the FAO, warned a forum in Rome marking World

---

[66]https://www.theguardian.com/commentisfree/2015/mar/25/treating-soil-like-dirt-fatal-mistake-human-life (accessed September 1, 2017).

[67]The examples relating to UK soils are likely representative of soils throughout Europe.

Soil Day in 2014.[68] The causes of soil destruction include chemical-heavy farming techniques, deforestation (which increases erosion) and global warming. The forum concluded that sustainability of the soil, which produces 95% of the world's food, is too often ignored by policy makers. The FAO cautions that unless new approaches are adopted, the global amount of arable and productive land per person in 2050 will be only one-quarter of what it was in 1960, due to soil degradation and population growth.

Soils have played a key role in absorbing carbon (CO_2) and filtering water. However, the FAO warns that soil degradation creates a vicious cycle, in which less carbon is stored because of decreased organic matter content, so the world gets hotter through the greenhouse effect and the land is further degraded. Volkert Engelsman, an activist with the International Federation of Organic Agriculture Movements, told the forum that 30 soccer fields of soil were being lost each minute, largely because of intensive farming.

Also under FAO auspices, a report in 2015 by 200 soil scientists from 60 countries, who had synthesized 2,000 peer-reviewed scientific publications over a period of 2 years, declared that soils of developed countries were often in a parlous state, as a result of industrial-level farming practices. Moreover, soils in developing countries are often in bad shape too. The 650-page publication "The status of the world's soil resources", produced by the FAO's Intergovernmental Technical Panel on Soils (ITPS), coincided with World Soil Day on December 4, and also the end of the UN International Year of Soils 2015, an initiative which has served to raise global awareness of what has been described as "humanity's silent ally".[69] The deterioration of global soil health was ascribed to an array of dysfunctions—erosion, compaction, acidification, contamination, sealing, salinization, waterlogging, nutrient imbalance (both nutrient deficiency and nutrient excess), and loss of soil organic carbon and soil biodiversity.

The ITPS Report prescribes what needs to be done to rescue the situation. For the world community to achieve sustainable soil management requires complex and multifaceted solutions. Implementation of soil management decisions is typically made locally and occurs over a wide range of socio-economic conditions. The development of specific measures appropriate

---

[68]https://www.scientificamerican.com/article/only-60-years-of-farming-left-if-soil-degradation-continues (article by Chris Arsenault, December 5, 2014) (accessed August 25, 2017).

[69]The Global Soil Partnership (GSP) was established in December 2012 by nation State members of the FAO, which Partnership then established the ITPS.

for adoption by local decision makers requires multilevel, interdisciplinary initiatives by many stakeholders. Partnerships are therefore essential.

(B) IAASTD

The International Assessment of Agricultural Knowledge, Science and Technology for Development (IAASTD) was initiated in 2002 by the World Bank and FAO, as a global consultative process. The objective of the IAASTD was to assess the impacts of past, present and future agricultural knowledge, science and technology on

- the reduction of hunger and poverty
- improvement of rural livelihoods and human health
- equitable, socially-, environmentally-, and economically-sustainable development

The outputs from this 4-year assessment were published in 2009, called "Agriculture at a crossroads", colloquially known as the World Agriculture Report. This comprised a Global and five Sub-Global reports; a Global and five Sub-Global Summaries for Decision Makers; and a cross-cutting Synthesis Report with an Executive Summary. The Summaries for Decision Makers and the Synthesis Report specifically provide options for action to governments, international agencies, academia, research organizations and other decision makers around the world. These reports draw on the work of some four hundred experts from all regions of the world, across a range of disciplines, who participated in the preparation and peer-review process. The IAASTD Executive Summary, while being approved by governments of 58 countries, did not receive the full backing of the other three governments—the United States, Canada, and Australia; this was seemingly related to the Report's stance on GMOs.

The IAASTD report acknowledges that demand for food will continue to rise as human population grows, and in the short-to-medium term, production is expected to increase to meet this demand. However, there is growing concern about the vulnerability of the productive capacity of many agroecosystems to stresses imposed by agricultural intensification, resulting for example in water scarcity and soil degradation (Thrupp, 1998; Conway, 1999; Millennium Ecosystem Assessment, 2005). Agricultural activities also account for 15% of global emission of greenhouse gas (methane, nitrous oxide and carbon dioxide), two-fifths of which is a result of land use or soil management practices (Baumert et al., 2005). The impact of nitrates from fertilizers and livestock production on soil and water resources is a related issue. This impact may be characterized as the nitrification of the global ecosystem resulting from inorganic fertilizers and alteration of the global nitrogen cycle. Eutrophication as a consequence of nutrient run-off from agriculture poses problems both for human health and the environment. Impacts of eutrophication have been easily discernible in some areas, such as the Mediterranean Sea and northwestern Gulf of Mexico (Wood et al., 2000).

The Report's specific findings on soil fertility were extensive. Soil degradation poses a considerable threat to sustainable growth of agricultural production and requires increased action at multiple levels. The direct influence of agricultural practices accounts for about one-quarter of total soil degradation (GACGC, 1994). Though inorganic nitrogen fertilizer has enabled crop production increases over the last 100 years, the downside has been that hitherto traditional knowledge-based and labor-intensive ways to maintain soil fertility have been put in abeyance. As a result, soils have become more acid while organic content has progressively declined, so that leaching of minerals has increased (requiring ever higher doses of fertilizers). Inorganic fertilizers have to a large extent sidelined the traditional practice of crop and pasture rotation, which enabled soil fertility to be maintained. Field-level livestock husbandry has declined and monocultures of grain crops have put their signature on the countryside.

The IAASTD report makes the case for an intensive relearning of this "old" knowledge for both industrial and small-scale agriculture, as well as its re-application in agricultural research. The report calls for refraining from all forms of agriculture and soil management that disregard the fundamental value of fertile soils. This includes overfertilization, overexploitation of sensitive soils, and exposure to water and wind erosion; the latter risk can be reduced using tree breaks or hedges. Other harmful practices include the use of heavy machinery which can lead to soil compaction, as can deep or unnecessary tillage with a plow.[70,71]

Since 1950, one-third of the world's soil has been profoundly altered from its natural ecosystem state because of moderate to severe soil degradation (Oldeman et al., 1991). Soil degradation has direct impacts on soil biodiversity, on the physical basis of plant growth and on soil and water quality. Processes of water and wind erosion, and of physical, chemical and biological degradation, are difficult to reverse and costly to control once they have progressed.

---

[70]http://www.fao.org/fileadmin/templates/est/Investment/Agriculture_at_a_Crossroads_Global_Report_IAASTD.pdf (accessed September 4, 2017).

[71]http://www.globalagriculture.org/report-topics/soil-fertility-and-erosion.html (accessed September 1, 2017).

The Global Assessment of Human-induced Soil Degradation (GLASOD) showed that soil degradation in one form or another occurs in virtually all countries of the world. About 2,000 million (2 billion) hectares are affected by soil degradation. Water and wind erosion accounted for 84% of this damage, most of which was the result of inappropriate land management in various agricultural systems, both subsistence and mechanized (Oldeman, 1994). GLASOD estimates that degradation has affected 38% of the world's cropland, to some extent as a result of human activity. An example given in the IAASTD report is that the intensive and continuous monoculture of rice-wheat systems in the Indo-Gangetic Plain of India and Pakistan has led to both soil and water degradation (Ali and Byerlee, 2002). Scientists have observed stagnating yields there in farmers' fields, despite increased use of inputs. These cropping systems, representing the "breadbasket" of northern India and Pakistan, cover 12 million hectares, on which the food security of some 500 million people depends.

Of the means to stem this degradation, the IAASTD report cites *inter alia* broader adoption of Conservation Agriculture practices. This would result in numerous environmental benefits such as decreased soil erosion and water loss due to run-off, decreased carbon dioxide emissions and higher carbon sequestration, reduced fuel consumption, increased water productivity, less flooding and more recharging of underground aquifers (World Bank, 2004).

Different soil types have very different erodibility characteristics, i.e., their ability to resist soil erosion caused by water, wind or plowing varies a great deal. Some soils will hardly recover once eroded, while others may regenerate within a relatively short time. There are two dimensions to the degradation of soils: first their *sensitivity* to factors causing degradation, and second their *resilience* to degradation, namely their ability to recover their original properties after degradation has occurred. Sensitivity and resilience depend on climate and the biophysical structures of the soil, and whether degradation has exceeded a threshold of resilience (such as loss of all organic matter or severe compaction) beyond which recovery is not possible without active intervention (Blaikie and Brookfield, 1987).

The IAASTD recognizes that in agriculture, there is most often a continuum between a farming system and a natural ecosystem, as the term "agroecosystem" indicates. Farmers have a pivotal role as managers of these systems, and as stewards of their resource base. Their role includes, for example, the conservation of soil properties and water availability, the development and maintenance of crop species, and the pursuit of multipurpose production objectives. Issues relating to Natural Resource Management (NRM) are often framed as specific problems such as soil degradation, water pollution and biodiversity loss. We should also frame agriculture's contribution to NRM positively: farmers create and enhance resources such as arable soil, agrobiodiversity and productive forest stands. Working with the natural resource base, farmers often enrich and enhance it.

(C) Edmondson et al. study in the United Kingdom

This research and discussion paper of 2014 led by scientists from the University of Sheffield relates to the deteriorating condition of intensively-farmed soils in the county of Leicestershire, United Kingdom (Edmondson et al., 2014). Field research showed that the natural capital of soils and the "ecosystem services" on which sustainable agricultural production is predicated are in retreat. In addition to lower soil nitrogen, the authors cite lower C:N ratio, loss of soil organic carbon (SOC) and greater soil compaction, which impede water holding capacity, water purification, drainage and flood mitigation measures. Soils in an urban environment are in better condition, however, leading the authors to advocate that more attention be paid in the United Kingdom to urban agricultural potential (home gardens, allotments, use of vacant land, etc.) to provide the needful increase in future food production. They point out that there are already over 800 million people involved in global urban agriculture, making an important contribution to global food security. In the face of deteriorating arable soil condition in the United Kingdom, Edmondson et al. opine that an agricultural crisis threatens, unless dramatic action is taken to reverse the depletion in arable soil nutrients, and urban agriculture is institutionalized as an additional/alternative source of food.

During the 20th century, the means to increase global food production and productivity on existing arable land entailed agricultural intensification. This led to more marginalized land being brought into cultivation, involving deforestation, with the cleared land then devoted to intensive crop or livestock rearing. Agricultural intensification has had an unwelcome effect on land globally, with reduced "natural capital", including SOC and mineral nutrient content. Loss of organic matter content is of particular concern for food security, as cereal crop yields usually increase linearly with SOC levels (Lal, 2004, 2010). A needful goal in support of sustainable agriculture and building resilience against food and livelihood insecurity on a global scale is to reverse the ubiquitous loss of SOC from arable land, and increase C:N ratios; the latter has a vital nutrient recycling role for recurrent cropping, and reduces the propensity for N leaching from the soil (Hopkins, 2008; Dungait et al., 2012; Robinson et al., 2013).

(D) UK Committee on Climate Change (CCC)

According to a report in 2015 by this UK government advisory body, huge swathes of British farmland will become unproductive within a generation, because of impoverished soil health and erosion, at a time when there is ever more need of food due to population increase in the country (CCC, 2015). Until now farmers have been able to increase yields using technologies generated by scientists. Lord Krebs, then chair of the CCC's adaptation subcommittee, cautioned that British farmers until now have been carefree with their use of soil, treating it as a mined resource which can be depleted, rather than as a stewarded resource that must be maintained for the long-term. The CCC Report lays the blame for decreasing fertility on intensive agricultural practices, such as deep plowing, short rotation periods, and exposed soil surfaces leading to soil erosion from wind and heavy rain. The report mentions that since 1850, Britain has lost 84% of its fertile top soil, with erosion rates of 1–3 cm a year; this is clearly unsustainable, as 1 cm of soil is thought to be 100 years in the making. Long-term remedial measures are foregone at the behest of vested interests in the farming industry; government policy makers are in denial of the impending crisis, being more interested in short-term benefits in a business-as-usual manner. A proposed European Parliament directive setting out rules to protect soils was withdrawn by the EU Commission in 2014, after several countries (Germany, France, the Netherlands, Austria and the United Kingdom) conspired to block the directive, arguing that there is sufficient regulation already.

(E) Environmental Audit Committee of the British House of Commons

In its report of 2016, the Committee warned that "Some of the most productive agricultural land in England is at risk of becoming unprofitable within a generation due to soil erosion and loss of organic carbon, and the natural environment will be seriously harmed" (UK Parliament, 2016). The Committee accepted that neglecting the health of "British" soil could lead to reduced food security, increased greenhouse gas emissions, greater flood risk, and damage to public health. While the UK government says it will ensure that all soils are managed sustainably by 2030, the Committee's inquiry suggests that the government's actions do not match its ambition, and casts doubt on whether the country is on track to achieve the 2030 goal. It also points out that knowing how farming activities are affecting soil health is crucial to developing effective policy, yet the United Kingdom lacks an ongoing national-scale monitoring scheme for soil health.

The Committee's lack of conviction that the government of the day is able to stand up to the vested interests of the conventional farming lobby is illustrated by an incident at the respected Oxford Farming Conference in 2015, at which DEFRA's Minister of the day, Liz Truss, displayed her remoteness from the soils crisis when a delegate waved the Edmondson et al. paper at her, insisting she acknowledge the scale of the crisis (Lymbery, 2017).[72] Her response was only to sidestep the issue, pointing to the richness of the black soil in her parliamentary constituency of the county of Norfolk. It was left to Professor Lord Krebs in his subsequent presentation at the Conference to point out that those Norfolk soils are deteriorating quickly and would not be there much longer if current trends and farming practices continue.

(F) Rickson et al. review paper, 2015

Global demand for food is increasing in terms of the quantity, quality and reliability of supplies (Rickson et al., 2015). Currently, over 90% of our food is grown on a non-renewable natural resource, the soil. This review examines the latest research on selected soil degradation processes (soil erosion by water, compaction, loss of organic matter, loss of soil biodiversity and soil contamination), and how they impact on food production. Every year, an estimated 12 million hectares of agricultural land are lost to soil degradation, adding to the billions of hectares that are already degraded. It is estimated that soil degradation leads to a potential loss of 20 million tons of grain per annum, but this is likely to be an underestimate. Comprehensive soil conservation practices are required to respond to the multiple problems of soil degradation if the world is to be able to feed more than 9 billion people by 2050.

11.1.2.2 Alternatives to Conventional Agriculture Which Could Sustain Soil Fertility

There is ample scope for food supply to be increased in sustainable ways using ecological agriculture based on traditional methods and more local control, as discussed above related to the Report by IAASTD (2009). Section 6.1 has also considered some of these. Monbiot (see Section 11.1.2.1) also says that we have solutions at hand, again citing Conservation Agriculture, which helps protect the soil from erosion through permanent soil cover. The practice of "Conservation Agriculture" in particular was discussed in Case Study 2 on the companion website of the current author's earlier book, shown to be a worthy and sustainable "post-modern" successor to industrial farming. The progressive decline of Soil Organic Carbon under intensive agriculture,

---

[72]DEFRA = The Department for Environment, Food and Rural Affairs, UK Government.

> **BOX 11.1 How to Increase Soil Organic Carbon**
> Peter Melchett of the UK's Soil Association, in his submission to the House of Commons Environmental Audit Committee (see above), said:
>
> *How you get carbon back into soil is fairly settled science, I would say: you use green cover crops in the winter, you do not leave the soil exposed, you use, where you can, crops with deeper roots so you have more biomass to put back into the soil. You put the crop waste, the straw, ideally through a cattle shed and then it is farmyard manure or compost back into the soil. You use rotations that include grass, exactly as you say, because that will help. It is not rocket science. What seems to be very difficult is to get Governments of all parties to recognise the problem and to recognise the need for action. It is becoming more urgent. The potential is huge, we could put huge amounts of carbon back into soil.*
>
> **(Melchett, 2016)**

and the associated dangers, have been cited above by the FAO, Edmondson et al. and Lal. Yet the situation is amenable to remedy (Box 11.1).

11.1.3 The Challenge of Maintaining Social Cohesion and Socio-Economic Integrity

In addition to alternative field husbandry and other agricultural practices to maintain soil fertility, some alternative food systems seek to improve the equity and social acceptability of the current mainstream model. These, *inter alia*, seek to shorten the distance, in both time and space, between producers and consumers.

Many forces give shape to the contemporary global landscape of food, downstream of those who grow it. These include multinational agribusinesses and supermarket chains, who process, package, distribute and market it, much of it in the developed world, and the international regulatory agencies, such as the World Trade Organization (WTO), which condition "free trade" in foodstuffs. "Free trade" has led to some unfortunate spin-offs, including environmental damage, production of processed foods of low nutritional value, the concentration of profit taking and exploitation of small-scale growers and laborers who work the land they do not own, and declining terms of trade for small-scale producers in developing countries which accompanied increase in trade volume.

Though the paradigm accompanying the formation of the WTO in 1995 was that a more commercialized agriculture and bolstering international trade in foodstuffs were the means to alleviate poverty in the global south, in practice this has not happened. Many studies have shown that land consolidation in the interests of efficiency and profitability has not led to the original smallholders being better off. On the contrary, those smallholders have been dispossessed of their land, which formed a major part of their social identity, just as much as the food and cuisine derived from it contributed to "identity". Klein and his colleagues from the London School of Oriental and African Studies (SOAS) have argued this case convincingly in their extensive review of literature (Klein et al., 2012).

Alternative foodways have evolved from the widespread disappointment with the scenario just described, with the intention of closing the gaps between food production and consumption, and better embedding society's cultural ownership of the land and the food harvested from it. The food sovereignty movement has its roots in such (see Section 6.8 in the current author's earlier book). Social anthropologists have laid claim to being able to provide special insights into how these alternatives may develop. For instance, anthropologists have looked critically at policy claims that traditional land tenure systems invariably impede higher production and sustainability, arguing instead that such systems often give farmers the flexibility and mobility necessary to cope with short-term crises (Agarwal, 1994; Breusers, 2001). In the anthropological literature covering this, the term "resilience" seems scarcely used, yet that is the issue at stake for the countless communities involved, in both developing and developed countries.

There is growing support for alternative food systems in the developed world, an article in the food and farming-themed issue of *Green World* magazine of Summer 2016 entitled "A slice of the good life", providing a window on it, in the shape of urban agriculture in London (echoing the emphasis of the Edmondson et al. paper mentioned above).[73] Another article in that magazine advocates making far more use of feeding human food waste to pigs.[74,75] Furthermore, several other highly pertinent articles relate to alternative foodways.[76,77,78]

---

[73]In Venezuela, there is a Ministry of Urban Agriculture.

[74]http://www.greenworld.org.uk/article/pig-idea (accessed August 15, 2017).

[75]Food waste was discussed in the current author's earlier book (Chapter 5.3.2). See also a video clip on the home-grown Robin Hood Army in Karachi, Pakistan, initiated by Sarah Afridi and others, redistributing edible "food waste" from the wealthy part of the city to slum dwellers. http://www.bbc.co.uk/news/av/world-asia-36247502/robin-hood-army-feeds-pakistan-poor (accessed September 25, 2017).

[76]http://www.greenworld.org.uk/article/local-food-movements-uk (accessed August 15, 2017).

[77]http://www.greenworld.org.uk/article/ethical-eating (accessed August 15, 2017).

[78]http://www.greenworld.org.uk/article/food-and-farming-introduction (accessed August 15, 2017).

Multiple benefits of "own-growing" include provision of healthy food (chemical- and GMO-free, if desired), improved diet, therapeutic psycho-logical stress-relief value of greater self-sufficiency, reduced "food miles" (sourcing of food over huge distances), and increased physical exercise/ healthier lifestyle. Not only does urban production espouse the "alternative" philosophy to conventional agriculture, but it reduces household expendi-ture on fresh food, which is hit by price inflation.

Social movements promoting alternative foodways have had varying success, reliant not only on the degree to which the movement is organized, but also on the cultural context in which they operate (McIntosh, 2013). For instance, the movement against GMO food has been more successful in arousing supportive public sentiment in the United Kingdom than in the United States (Schurman and Munro, 2009). The argument used in the United Kingdom was replete with damning language against the practice, such as "unsafe, unnatural and associated with unknowable and uncontrol-lable public and environmental risks", and set against the perceived arch-enemy, the US company Monsanto. It seems that the anti-GMO campaign in the United States was less effective as the American consumer culture more readily adapted to GMO foods, since the public there had more trust in business. Moreover, the anti-GMO movement in the UK directed at US farmers which grew the GM crops and companies which facilitated them to do so and marketed the same, created a backlash within the American public, as it was perceived to be anti-American.

Cleveland (2014, op. cit.) argues that combining selected aspects of small-scale traditional agriculture with modern scientific agriculture can help bal-ance our biological need for food with a lower environmental impact, and nurture our cultural, social and psychological needs related to food.

In a challenging landmark book, McKeon (2015) opines that food se-curity of the majority in the world is held to ransom by the power of food corporations and power blocks which connive together in an exploitative way which, together with land grabs for intensive farming, results in small-holders being evicted from land they have used under slash-and-burn tech-niques for generations. She advocates for a re-balancing of the global food system which has evolved since the Second World War, which currently discriminates against the interests of smallholder farmers, pastoralists and fishers, who produce most of the global food, in favor of mega-scale com-mercial interests. She is a strong advocate of food sovereignty and a rights-based approach, and laments the apparent demise of democratic debate around food issues.

McKeon details the rise in community-based ways of farming and marketing, which underwrite rather than destroy the connection between people and their livelihoods, the land they farm, and the food they eat. She has hope that the Committee on World Food Security, revamped in 2009, can assist this rescue through policy initiatives. She calls for effective public policy instruments in support of accountable governance, and for the untrammeled power of global markets and corporations to be regulated. This would create a more level playing field such that a more equitable food system evolves and flourishes, this being more resilient to natural resource shocks and providing affordable food in a sustainable way. Currently, the economic equation rides roughshod over the concept of hunger, resource depletion, environmental degradation, and not least, human rights for those at the bottom of the pecking order.

McKeon's treatise is a huge and timely socio-economic and equity challenge to the *status quo* of the current industrial farming-dominated global food system, effectively run by big-business interest groups. The scenario she portrays is powerfully summarized in a review article in the *Journal of Peasant Studies*, which clearly espouses McKeon's thesis (McMichael, 2015). McKeon argues that the global food system is market-centered and market-driven, with little systematic prioritization of hunger and undernutrition, ecosystem degradation and resource depletion, and producers' and workers' rights. Governance of the existing system prioritizes commercial private sector interests and the global public sector and development agencies have become hi-jacked by these, playing "second fiddle" to them.[79] Smallholder and local community subfood systems have been marginalized and brow-beaten into conforming with "big brother" interests.

Yet, as mentioned above, these smallholders, pastoralists, fishers and urban gardeners produce more than 70% of the world's food. The corporate food system is therefore undemocratic by default, argues McKeon. Power currently does not lie within these micro-systems but with corporate interests which control food chains, high-investment marketing and GMO development, which consolidate property relations while proclaiming market freedoms, an obscene contradiction McKeon maintains. She draws attention to how different this is from the local food systems espoused by Green Movements globally, and the actions of cities like Belo Horizonte in Brazil, which since 1993 has become a world pioneer in delivering local

[79]Sridhar drew attention to one such development agency wielding power at the expense of local communities in India (see Chapter 6.2.2), in respect to nutritional security.

governance in food security matters, with a municipal law which sets a policy framework committed to the concept of food sovereignty.[80] McKeon maintains that the consequences of current food geopolitics have impoverished developing countries, rendering them increasingly vulnerable to global forces beyond their control, thereby lowering their resilience against food insecurity. She argues for the building of effective public policy instruments in support of global food democracy, which empower accountable rights-based governance and curb the power of markets and corporations.

11.2 CASE STUDY 2. RESILIENCE TO FOOD AND NUTRITION SECURITY AMONG THE INUIT[81]

11.2.1 Introduction

This case study is a continuation of Section 8.3. The Inuit comprise an indigenous people of northern Canada, and parts of Greenland and Alaska. In his book "The Other Side of Eden", anthropologist Hugh Brody expressed the need for the hunter-gatherer world view to be preserved and recorded, at a time when our "natural environment" is so significantly under threat (Brody, 2000). Whereas Brody could not find his way in the Arctic without his Inuit guide, in his book he tells of his guide visiting London and unaided unable to find his way among the "cliffs". This case study is intended to bring something of the Inuit cultural and nutritional ways and means to those who are focused on the "developing world", and who would perhaps not otherwise venture there. The current author is of the opinion that there is much that the latter group (including himself) can learn from the traditional lifestyle of the Inuit, with whom he has never worked, and in the process come to respect their tenacity and resilience—another case of applying and benefitting from some "lateral thinking" (Chapter 9).

The indigenous people of Canada refer to themselves under the term First Nations or Native People. This group includes the Inuit, Dene, Métis and many distinct Amerindian cultures, such as the Dakota, Cree and Haida (INAC, 2015). The Canadian Encyclopedia describes the Inuit as an aboriginal people, the majority of whom inhabit the northern regions of Canada.[82] An Inuit person is known as an Inuk. The Inuit are not an homogeneous

[80]http://www.futurepolicy.org/food-and-water/belo-horizontes-food-security-policy (accessed September 5, 2017).
[81]The term Inuit used here refers to people of the northern hemisphere whom the unenlightened may still call "Eskimo"; the latter term is now considered offensive and demeaning.
[82]http://www.thecanadianencyclopedia.ca/en/article/inuit (accessed August 29, 2015).

group, rather comprising eight main ethnic groups. In the vernacular language (Inuktitut), Inuit means "the people". The language has five main dialects in Canada. The Inuit homeland is known as Inuit Nunangat, which refers to the land, water and ice contained in the Arctic region. The term Inuit Nunangat may also be used to refer to land occupied by the Inuit in Alaska and Greenland.

In 2011, using data from the National Household Survey, Statistics Canada estimated that 59,440 people in Canada identified as Inuit. In 2011, ~73% of all Inuit in Canada lived in Inuit Nunangat, with nearly half living in Nunavut, followed by Nunavik (in northern Quebec), Nunatsiavut (located along the northern coast of Labrador) and the western arctic (Northwest Territories and Yukon), known as Inuvialuit.

Nunavut, with its administrative and commercial capital of Iqaluit on the southern part of Baffin Island, had a total population of some 32,000 people (in 2011), across 808,185 square miles. The Inuit comprise more than 80% of Nunavut's population, the remainder being of European ancestry. Owing to the high fertility of the Inuit, this percentage is increasing, with one-third of the population being below 15 years of age.

Nunavut is a huge sparsely-populated territory of northern Canada, stretching across most of the Canadian Arctic archipelago, comprising mountains and tundra with permafrost. It lies above the northern tree line, which has a NW-SE alignment; this line approximates to the traditional cultural boundary between the Inuit to the north and North American (First Nations) Indians, known as the Dene, to the south. The vegetation is of stunted hardy shrubs (especially dwarf birch), grasses, flowering plants, mosses and lichens. Herbivorous mammals feed on this cover—caribou (reindeer), muskox and smaller mammals. The predator level of the food chain includes red and Arctic foxes, wolves and grizzly bears. Marine-based mammals at the coast comprise polar bear, walrus and various types of seal, with beluga, bowhead whales and narwhals frequenting coastal waters. Migratory aquatic birds (such as species of ducks and geese) nest during the summer, though only a few species live there year round.

According to the Canadian Encyclopedia, traditionally the Inuit have been hunters and gatherers who move seasonally from one camp to another. Large regional groupings were loosely separated into smaller seasonal groups, winter camps (called "bands") of around 100 people and summer hunting groups of fewer than a dozen. Each band became roughly identified with a locale and named accordingly—for instance, the Arvirtuurmiut of Boothia Peninsula are called "baleen whale-eating people".

11.2.2 Food Security Strategies

11.2.2.1 Inuit Hunter-Gatherer Communities

Hunting and gathering of "country food", harvested from both the marine and land environments, has traditionally been the way of sustenance for the Inuit (whether in Canada, Alaska or Greenland), passed down from one generation to the next, through word of mouth and actions. There would be no escape from it for the youngsters, hearing the vernacular language in the home and participating in the traditional lifestyle throughout the year. As soon as the youth were capable, they would accompany their parents and other family members on hunting expeditions or collect plant food during the warmer months. There is a distinct gender division of labor, with the men involved with the hunting and fishing, the more dangerous pursuits, while the women process, preserve and cook the meat and fish, and collect and process the wild plants and fruits of summer. This lifestyle has served the Inuit well over centuries if not millennia, amply providing their food and nutrition security, until recently. The main food security challenge for the traditional villagers living away from the sea has been avoiding starvation in late winter if primary meat sources become too scarce or lean.

Hunting and gathering has been not just a necessary survival pursuit, but has equated to "identity" too, just as much as has the shared language and art, family cultural norms, attitudes and behavior (see Box 11.2). Good hunters are respected, as is a good work ethic—lazy people who fail to contribute to the community are not. They are just more mouths to feed in a place where food can be very hard to come by. There is total bonding between the people and the environment. In Inuit traditional culture the connectivity between humans, animals, plants, the land they live on and the air they share has been ingrained from birth. In industrial societies, people do not speak familiarly about "our" food animals, as happens among

BOX 11.2 The View From Alaska

As expressed by Patricia Cochran, an Inuit from Northern Alaska (an Inupiat):

You truthfully can't separate the way we get our food from the way we live. How we get our food is intrinsic to our culture. It's how we pass on our values and knowledge to the young. When you go out with your aunts and uncles to hunt or to gather, you learn to smell the air, watch the wind, understand the way the ice moves, know the land. You get to know where to pick which plant, and what animal to take.

BOX 11.2 The View From Alaska *(Cont'd)*

It's part, too, of your development as a person. You share food with your community. You show respect to your elders by offering them the first catch. You give thanks to the animal that gave up its life for your sustenance. So you get all the physical activity of harvesting your own food, all the social activity of sharing[83] and preparing it, and all the spiritual aspects as well. You certainly don't get all that, do you, when you buy pre-packaged food from a store?

(Gadsby, 2004)

Cochran directs the Alaska Native Science Commission in Anchorage, which promotes research on native cultures, and the health and environmental issues that affect them.

traditional Inuit, who have a sense of kinship with food sources. There is no single word for snow or seal, for instance, but rather precise words to describe which *type* of snow or seal. Such linguistic diversity can have come about only over a long period, a product of an intimate connection of the people and their habitat.

The traditional lifestyle has been a provider not only of food, but also of heat, light, clothing, tools and shelter.[84] The skin of some of the mammal kills is used for clothing, blankets, tents and boats. When food is cooked prior to eating, seal oil is used for the cooking and also for lamps, while bones, ivory and wood are used to make tools. Little is wasted, there is no pollution, and animals and people lived in harmony with a land that most people from lower latitudes of North America, say, would find hostile and intimidating.

11.2.2.2 Cultural Adaptation of Modern Inuit

The traditional hunter-gather life of the Inuit has been rewarded by both food and nutritional security, though it is associated with low life expectancy. The latter seems not to be caused by food and nutrition insecurity, but linked with the dangers men faced during the hunt for wildlife and fishing, the over-crowded and unhygienic and unsanitary traditional homesteads, and the dearth of health services in remote villages. Yet times for the Inuit have

[83]Originally, "reciprocal sharing" was believed by anthropologists to condition the means of food distribution, on a demand or offering basis. This would constitute a risk-averse strategy whereby food is considered a club good rather than being owned solely by the family or group which made the kill. Over the last decade though, this understanding by outsiders has been disaggregated (Kishigami, 2004).

[84]http://www.arcticvoice.org/inuit.html.

changed, since visitors from the outside world started arriving at their remote communities, bringing with them new opportunities and dangers. The traditional way of life as described above is now rarely found; it has been replaced by a way of life which is very different, though not completely so.

The Inuit first encountered Europeans shortly after the Norseman Eric the Red colonized the west coast of Greenland in around AD 1000. Following that, the British adventurer Martin Frobisher made contact with the Inuit in 1576, and after him came missionaries, then whalers in the 19th century, explorers, miners of oil, gas, diamonds and iron, scientists, traders (in furs, skins, fish and shrimp), and more recently those involved in the tourism industry. Many miners and traders set up bases in traditional Inuit territory, which developed into "new" settlements, often in clusters. The Inuit came to depend on income from casual employment to purchase items which became "necessities". Very few Inuit now subsist solely in the traditional way on hunting, trapping, fishing and gathering.[85] To explain this, there were (and continue to be) "push" factors impacting traditional Inuit communities, as well as "pull" factors; these are discussed below.

Contact between First Nations and outsiders has brought about changes at both group and individual levels. The process of psychological acculturation gives rise to new behaviors and strategies of individuals and communities concerned (Berry, 1997). Croal and Darou (2002) provide insights (for outsiders) on the attitude of First Nation peoples to "international development", to reduce poverty, for example.

Concerning the "push" factors, nature's largesse from both land and sea has become even less reliable than ever it was, for several reasons. One reason is climate change. Arctic average temperatures have already risen at about twice the rate of the rest of the world over the past century and this is likely to increase up to three times the global temperature rise over the next 100 years. The World Meteorological Organization (2017) states that the extent of Arctic sea-ice in 2016 was well below average. The reduction in reflective snow and sea-ice cover reveals darker land and sea surfaces which absorb heat from the sun, thereby enhancing global warming (Box 11.3). Traditional villages at the edge of the ocean are being threatened.

Related to the sea level rise is the melting of the permafrost on which the traditional "old" settlements are built, causing them to destabilize and sink. The winter sea-ice, which has protected these ocean fringe villages

---

[85]The same is true for most of the world's traditional hunter-gatherer communities, for almost all have come into contact with the modern world, for better or worse.

BOX 11.3 The View from the IPCC

... the annual mean Arctic sea-ice extent decreased over the period 1979 to 2012, with a rate that was very likely in the range of 3.5 to 4.1% per decade. Arctic sea-ice extent has decreased in every season and in every successive decade since 1979 ...

(IPCC, 2014)

from ocean swells and storms, has become less thick and breaks up earlier in the spring, thus making the villages more vulnerable to those swells and storms. The thinning of the ice also means that it becomes more dangerous to cut fishing holes in the ice over deeper water, where the fish shoals are located. The Inuit are experiencing first-hand the adverse effects of climate change (Box 11.4).

In the interior, where there were traditional settlements too, the southern boreal forest is creeping north as the temperature rises, replacing the tundra grazing grounds of the caribou. This has had a marked effect on caribou migration patterns in living memory, and the quarry is now scarcer than it used to be.

As well as these "push" factors, there are "pull" factors at play on the traditional Inuit, in the form of large modern settlements, initially set up as trading posts for the oil, mining and whaling industries. As the uncertainties of hunting and gathering have multiplied, and with them increased vulnerability to food insecurity, so more and more Inuit have drifted to these settlements, to participate in the cash economy on offer. Through providing their labor, albeit cheaply, they have been able to purchase food and other items from the stores there. Government has played a large part in laying on services and social protection for these settlements, adding to the attraction.

BOX 11.4 The View From Greenland

... the Arctic is the barometer of the globe's environmental health. You can take the pulse of the world in the Arctic. Inuit, the people who live farther north than anyone else, are the canary in the global coal mine ...

Aqqaluk Lynge, leader of Greenland's Inuit population and former President of the Inuit Circumpolar Council; extract from submission to the Stansted airport (United Kingdom) inquiry (as reported in the *Independent* newspaper, May 30, 2007).

A greater dependency has developed for goods and services sourced in distant places, available only by sea or air transport.

The combined effect of these push and pull factors has persuaded the Inuit to opt for the settled condition, in order to improve their resilience against food insecurity (without the concept being framed as such by them). Faced with the decreasing bounty of food from the wild, the prospect of supplementing this with food from food stores in settlement towns has appeared attractive; such was the result of surveys done among the new settlers addressing the question "Why did you move?".

Yet much of the money that the new settlers earn goes towards buying the "modern" food which is sold there, the high prices reflecting the long-distance that that food needs to travel by air or sea from areas of production. Those new settlements are not as remote as the villages on the ocean rim, though they are still remote from the markets in south Canada or beyond that, usually with no road or rail access.

But the new settlers hardly ever rely solely on these expensive processed or frozen foods off the shelves or in the deep freezers of settlement stores. To retain their Inuit identity, most ensure they source "country food" in one way or the other, either by hunting and fishing intermittently themselves, or by purchasing from the remaining traditional hunter-gatherers. Yet hunting and fishing from a modern settlement base has also become expensive, with rifles and snowmobiles taking the place of husky dogs, sleds and skis, and modern boats with outboard engines replacing kayaks and spears. These new tools-of-the-trade do not come cheap, as they have to be flown or shipped in, together with the recurrent costs of ammunition and fuel, and maintenance. So economic access to food, and food of choice, is far from assured in the new settlements, especially for those who do not find work there, and live as dependants with relatives or friends.

Wenzel (2013) in a highly readable review article seeks to explain whether previous decades of research on the socio-economy of Inuit society, embedded in a voluminous literature, has shed much light on the theoretical anthropological issues involved. Wenzel concluded that research on Inuit economy has, in fact, contributed importantly to a rethinking of the shape and content of what "subsistence" means. Formerly considered as encompassing the most basic economic activities, the evolution of both term and practice over a relatively short period of time is now understood as a *cultural adaptation*. It is not the case that the "traditional" way of securing nutritious food has been usurped and replaced (as has been discussed above). Wenzel concluded that this was important because few hunter-gatherer societies today can be characterized as they were at the time of the "Man the Hunter" symposium in 1966 (Lee and DeVore, 1968). Today, hunter-gatherers, from the Arctic to Australia experience near-constant contact with market economies and a reality in which money plays a critical part in their livelihoods. It is in this regard that research on Inuit, as noted by Sahlins (1999), has conceptually contributed both to hunter-gatherer studies and to anthropology. As Wenzel points out, the dilemma confronting Inuit and, increasingly, all hunter-gatherer subsistence cultures is neither the penetration of money *per se* into its milieu nor its increasing necessity for traditional food gathering. Rather, it is that the production of money requires a very different pattern of labor allocation than many traditional resource activities—principally, its acquisition is through the sale of the individual's labor rather than through collective action.

Today the Inuit have adapted to the changes brought by the West and it is not uncommon to see an Inuit fishing through a hole in the ice while talking on a mobile phone, or drinking a can of fizzy drink as he drives his dog sledge. However, the Inuit hunting tradition remains and, particularly further north, is practiced with pride. In many areas the local people still hunt, fish and trap, and rely on their environment for food. The President of the Inuit Circumpolar Council, Greenland, Aqqaluk Lynge, has said: "Eating what we hunt is at the core of what it means to be Inuit. When we can no longer hunt on the sea-ice, and eat what we hunt, we will no longer exist as a people". [86]

11.2.3 Food and Nutritional Security of the Traditional Lifestyle Compared With That in the Modern Settlement

Wildlife has comprised the Inuit's diet since the people came to the area hundreds of years previously, and until recently they have never needed to buy food. According to Inuit testimonies recorded on a video documentary of 2015 by Mark Andrew Boyer, one lady says that 70% of people in Nunavut are food insecure, with 7 out of 10 pre-schoolers living in food-insecure homes. [87] A male informant reveals he has just one or two meals a day, sacrificing to feed his children. "People are hungry", says Nunavut resident Leesee Papatsie (creator of the Feeding My Family Facebook group)

---

[86] http://www.arcticvoice.org/inuit.html (accessed August 29, 2015) (quoted with permission).

[87] http://www.theatlantic.com/video/index/394583/nunavut-hunter-gatherer-society-runs-out-food/, subtitled "Feeding Nunavut: what happens when a hunter-gathering society runs out of food?" (accessed August 29, 2015).

in this documentary. Hunting has become a part-time activity for settlers while the number of "full-time" hunters in the region continually falls.[88] Nunavut has the highest food insecurity by far in Canada, according to a report in May 2017 by the Conference Board of Canada, with 51% of the Inuit population in Nunavut suffering from moderate to severe food insecurity (Conference Board of Canada, 2017). The Report's authors classify "moderate" food insecurity as being where someone reduces the amount or quality of the food he or she is eating; "severe" insecurity involves someone making major changes to his or her diet.

Associated with the relocation of most Inuit into new settlements, nutritional security is in decline.[89] This has given rise to the "Inuit Paradox" (Gadsby, 2004) whereby the nutritional benefits of the traditional Inuit diet has become evident, compared with the nutritional health of individuals in the new settlements, and at lower latitudes in North America, say (especially of individuals eating a "Westernized" fast food diet). In her excellent article, Patricia Gadsby considers how the traditional diet fulfils the Inuit's nutritional needs, when the food intake is high in protein and especially high in fat, with rather little in the way of fruit and vegetables. Nutrition gurus in "the West" say that too much fat is bad for health, and would not the deficiency in fruit and green vegetables in the traditional Inuit diet for most of the year lead to deficiency of Vitamin C?

Gadsby examines the traditional diet by interviewing nutrition researchers and descendants of those close-knit hunter-gatherer communities whose families still led a more or less traditional way of life (Box 11.5).

Arctic char (of the salmon family), seal, walrus and whale comprised the main provender from the sea, and from the land there was caribou, moose and large birds. Nuts, berries, roots, leaves, flowers and sorrel grass were harvested from the land in summer, and seaweed from the coast. There was protein in plenty from this diet, though protein cannot be the sole source of energy for humans if carbohydrate is in short supply. If a diet is too rich in protein, then the human body can waste away, just as it does if carbohydrate is insufficient—research has shown that anyone eating a meaty diet that is low in carbohydrates must have fat as well (Stefansson, 1935).

---

[88]http://www.arcticphoto.co.uk/inuitcan.asp (accessed August 29, 2015).

[89]There is a parallel here with new settlements for "indigenous peoples" in the Australian interior, where poor-quality processed foods seem easier to come by than wholesome equivalents at community stores, this impacting nutritional status and health outcomes. The government of Australia is on the case, to close the gap on indigenous disadvantage among the 80,000 or so people involved (see Australian National Audit Office, 2014).

BOX 11.5 The Inuit Traditional Diet: We Are What We Eat

One of the informants for the Gadsby article was Patricia Cochran, cited earlier, who spoke of the native foods of her childhood:

We pretty much had a subsistence way of life. Our food supply was right outside our front door. We did our hunting and foraging on the Seward Peninsula and along the Bering Sea. Our meat was seal and walrus, marine mammals that live in cold water and have lots of fat. We used seal oil for our cooking and as a dipping sauce for food. We had moose, caribou, and reindeer. We hunted ducks, geese, and little land birds like quail, called ptarmigan. We caught crab and lots of fish—salmon, whitefish, tomcod, pike, and char. Our fish were cooked, dried, smoked, or frozen. We ate frozen raw whitefish, sliced thin. That's why some of us here in Anchorage are working to protect what's ours, so that others can continue to live back home in the villages, because if we don't take care of our food, it won't be there for us in the future. And if we lose our foods, we lose who we are. The word Inupiat means "the real people". That's who we are.

Vitamin C was in plentiful supply too, from whale skin and caribou liver, seal brain, seaweed (kelp), and fruit harvested in summer. Moreover, pre-digested vegetation in the slaughtered caribou's stomachs was eaten too, this vegetation being too coarse for humans to digest first themselves—certainly traditional Inuit were/are not falling apart with scurvy, the related deficiency disease. Traditional Inuit practices like freezing meat and fish and frequently eating them raw conserve vitamin C, which is easily lost in cooking. The entrails of mammals are a rich source of nutrients too; seal livers are rich in vitamin D, for instance.[90] A rich source of (oil-soluble) vitamin A is seal oil, as are the oils of cold-water fishes and sea mammals, as well as animals' livers.

The meat of grain-fed livestock, which fills supermarket freezers, contains solid, highly saturated fat. In the case of wild animals though, which range freely and eat what nature intended, less of their fat is saturated, and more of it is in the monounsaturated form (like olive oil). That "not all animal fats are created equal" is at the heart of the "Inuit paradox". Gadsby (2004) quotes Eric Dewailly, a professor of preventive medicine at Laval University in Quebec. In the Nunavik villages in northern Quebec, adults over 40 get almost half their calories from local foods, says Dewailly, and the villager's cardiac death rate is about half of other Canadians or Americans.

---

[90] In lower latitudes, the human body can make Vitamin D using sunlight on the skin; this source is not an option for Inuit living at high latitudes, especially in winter.

The diet of traditional Inuit was also awash with omega-3 fatty acids (polyunsaturated fats), which cannot be synthesized by the body, and which, among other benefits conferred, are believed to protect the heart from life-threatening arrhythmias that can lead to sudden cardiac death (Surette, 2008). Cold-water fishes and sea mammals are particularly rich in these nutritionally more beneficial omega-3 fatty acids, while the polyunsaturated fats in the diets of most Americans are mainly the omega-6 fatty acids richly supplied by many types of vegetable oils (such as soy, sunflower and corn).

11.2.4 Overarching Inuit Cultural Factors Affecting Resilience

In traditional Inuit society, elders guide the youth in behavioral patterns by example, and this is reinforced by peer group pressure to conform, thereby deserving the respect of elders. In the new settlements, however, a communication gap has come about, due to the schools there teaching in English, and to a lesser extent in French, whereas the Inuit languages are not a medium of instruction. Children grow up speaking these new languages, rather than the traditional vernaculars. Conversely, the elders do not learn these two "new" languages, relying solely on their vernacular. Wisdom can no longer, therefore, easily be passed from elders to the youth, including knowledge on food and nutritional security. The older generation is becoming increasingly distant from the younger one, unable to communicate with it. Instead, the elders source their "news" not from radio, TV and the internet as do the young ones, but from community groups specifically for older people, using Inuit languages.[91]

Moreover, there is insufficient knowledge within the new settlements of nutritional matters in respect of the "new" foods and drinks available in the new settlements. In the old way of life, as discussed above, the communities' nutritional needs were met, the result of feeding practices entrenched over centuries—the communities knew that they would be well-fed and nourished if they continued the ways of the ancestors.

The Inuit's resilience to food and nutrition insecurity in the traditional lifestyle has changed downstream of the cultural adaptation to the lifestyle in these larger settlements. Life expectancy has increased, as a result of fewer deaths of active males in hunting or fishing accidents, and improved hygiene and shelter conditions, and medical and educational facilities. Also, the constant battle to secure sufficient food from nature has largely been removed. However, many Inuit in the new settlements are hungry due to

[91]https://video.nationalgeographic.com/video/exploreorg/inuit-wisdom-eorg.

economic access constraints, and nutritional security has tumbled. The decline is the result of a significant portion of the settler's food comprising nutritionally-poor processed and packaged food and drink, and meat produced in intensive grain-fed livestock systems, which have replaced much of the "country food". As a result, some common "Western diseases" are noted, such as Type 2 diabetes and obesity, and rotten teeth. The change in lifestyle has also led to significant alcohol and substance abuse, delinquency and crime—associated, it seems, with a certain longing by those involved for a return to the traditional lifestyle. The young are at risk through their adopting the new lifestyle without understanding the need for caution and self-discipline, while the old are feeling ill at ease with the new lifestyle and ostracized by it.

The predominantly Inuit region of Nunavik needs to address these major challenges and opportunities, through developing options for more employment and wealth creation in a culturally-sensitive way. The need for this is well-recognized by both the indigenous population and the Canadian government. The Inuit are expressing their views so they do not become sidelined in the exploitation of the region's natural resources. To this end there was an economic forum in Kuujjuaq, Nunavik in April 2010, to provide an opportunity for regional and local stakeholders to obtain information on specific economic opportunities for Nunavik and to discuss their merit for the communities. The possibilities resulting from this forum to provide economic and social stability and a sustainable development strategy for Nunavik are discussed in a paper published in 2014 in the journal *Polar Record* (Rodon and Schott, 2014).

The time has certainly come to move beyond learned scientific syntheses such as that of 2014 just cited, and that of the Canadian Conference Board of May 2017, to which reference was made above. The need now is for government policy updates, with matching costed-action plans implemented, the latter guided by Inuit society, the leaders of which given the opportunity to take ownership of that implementation. This should be the way forward for Canada, to prevent Nunavik from unraveling in terms of the socio-economy and the fabric of Inuit society. There are big plans under way to develop the mining industry, yet these need to be matched with progress on the social side too. Otherwise, the same may happen in Nunavik as happened in Liberia, as a result of the land use concessions awarded, growth occurring without development (Clower et al., 1966).

So the situation in "transition" Nunavik is, as was reasoned many a time earlier in the book, for the "developing world"—there needs to be a social side in development initiatives, not just technical advances. And the social concerns and priorities cannot be as junior partner or "country cousin" to technical ones, but rather on an equal footing … maybe more than equal!

11.3 CASE STUDY 3. HUMAN CAPITAL AS A RESILIENCE STRATEGY IN THE PAMIRS

11.3.1 Political and Economic Context

Following the introduction of *glasnost* and *perestroika* by Mr. Gorbachev in the mid-1980s, one Soviet republic after another peeled away to establish its own national sovereignty. Tajikistan declared such in August 1990. Political parties and allegiances were formed and tested, yet this resulted in ethnic clashes in 1992 and 1993, which comprised a civil war, followed by 7 years of instability and insecurity. Up to 100,000 people were killed and there were many atrocities and massive population displacement. Associated with the humanitarian disaster was the collapse of the Tajik economy. The majority of the population in Tajikistan came to live at or below the poverty line, and the country fell further and further down the UN Human Development Index. Pensions became insufficient to buy the bare necessities, especially in the face of rampant inflation.

This case study relates mainly to the remote mountains of Gorno-Badakhshan of eastern Tajikistan, based on the exhilarating ethnography of the region by Frank Bliss (2006).[92] The degree of resilience against food and nutrition insecurity is seasonally-determined; the biggest challenge is during the winter season. Bliss records that, prior to the Soviet occupation, most deaths over a year occurred towards the end of winter, resulting from seasonal afflictions of colds and chills. As the winter season progressed, and sources of dry fuelwood became scarce, it was increasingly difficult for the people to keep warm and cook food. Food and nutrition security was also compromised by decreasing availability of livestock fodder, which led to ever-reducing amounts of milk being produced.

After 1991 and the withdrawal of food and coal subsidies, the repatriation of Soviet specialists (in the energy sector, agriculture and public sector governance, for example), and the scarcity of vehicle fuel and other goods, the earlier Pre-Soviet occupation pressures on survival as described above were reinstated. A massive collateral effect of the Soviet collapse for people

[92]Since Olufsen and Schulz published their monographs on the Pamirs in 1904 and 1914, respectively, this was the first book to deal with the history, anthropology and recent social and economic development of the Pamiri people in Gorno-Badakhshan.

of the Soviet satellite States, with Gorno-Badakhshan being a case in point, was that funding of public sector infrastructure and services was hugely reduced. Pamiri community income and living conditions fell away to the level of a poor Sahelian country. Public sector staff including erstwhile scientists, university professors and teachers were unable to sustain the urban lifestyle to which they had become accustomed. Many had little option but to adopt ancestral subsistence agriculture as a means of providing food for their families, using animal traction and rudimentary equipment to plant potatoes and wheat as their staple foods. Some aspects of resilience against food insecurity in post-Soviet Tajikistan are discussed below.

11.3.2 Educated Cadre of First-Generation Farmers

A key difference between the ancestral times and those pertaining in 1991 was that this new breed of farmers and livestock keepers was not the uneducated cadre of old. They had been educated during the Soviet era, and had good numeracy and literacy skills. This proved to be a huge human capital asset for survival and indeed an excellent resource for economic recovery.[93,94] The first sign of this was an increase in agricultural production, something that had never occurred during Soviet times.

There were teething problems during the transition period though, a downside of the high educational status of the new farmers. The artisanal class—blacksmiths and carpenters, for instance—were underrepresented, as were those who could devise and operate simple irrigation equipment once the State Farm irrigation systems had failed through lack of spare parts. With the supply chains broken after 1991, people had to make their own footwear, for example, learning manual leather-working skills that were and still are common in the rural bazaars of southern Asia.

Though private sector entrepreneurial-cum-management skills had not been significantly developed under the communist regime, there were nonetheless elementary skills born of informal trading to "make ends meet", and these needed to be, and were, developed further post-1991. As a result, a better resilience against the threat of food insecurity was born.

---

[93]Human capital is the stock of knowledge, habits, social and personality attributes, including creativity, embodied in the ability to perform labor so as to produce economic value. http://www.oecd.org/insights/humancapitalhowwhatyouknowshapesyourlife.htm#Summaries.

[94]See also a World Bank Report of September 18, 2015: "Transitioning to better jobs in the Kyrgyz Republic: a jobs diagnostic". http://documents.worldbank.org/curated/en/781941468197944136/pdf/99777-REVISED-WP-P114024-P104706-PUBLIC-Box394824B-JD-Revised-8-122115-ENG.pdf.

The comparison with a Sahelian country made above, however, needs elaboration. In some ways the post-1991 situation in Gorno-Badakhshan was worse than the Sahel, for in the latter (apart from periods of drought and famine) there is normally some food, such that economic access to it, rather than its availability, is the challenge. Yet in post-1991 Gorno-Badakhshan, food was both in short supply *and* economically inaccessible. Moreover, the residents of Gorno-Badakhshan had to cope with a long and bitter winter in which agricultural and livestock fodder production was negligible. Nor were the people conditioned by experience to being plunged into a crisis threatening their very survival, and they had neither the institutional strategy nor the technical skills to cope with or escape from such a food security crisis. Yet, as mentioned above, on the positive side, the potential for recovery from crisis in Gorno-Badakhshan was higher than in a Sahelian country because of both the better educational level of the people *and* the adequate rainfall.

There had been unfortunate decisions made by Soviet planners which added to the challenges faced by Gorno-Badakhshan following the post-1991 economic collapse, reducing community resilience against food insecurity. For instance, many high-potential temperate fruit orchards (apricot, apple, mulberry and grape) had been cut as a result of central planners opting to replace them with the livestock fodder alfalfa. As a result of this, together with the poor-quality management of the remaining trees, fruit production in the early 1990s fell to less than 10% of what it had been 50 years earlier.

After the Soviet collapse, because of the geographical remoteness of Gorno-Badakhshan, it and other high mountain areas were more or less forgotten compared with lower-lying land in the country, with its less severe weather. Famine in the Pamirs was officially acknowledged for the first time in January 1993. The mountain people of Gorno-Badakhshan would have suffered even more from their isolation had not an Aga Khan Development Network (AKDN) project in 1993–94 provided broader support. Bliss (2006) opines that without that substantial humanitarian assistance, by 1994/95 there would have been a food supply crisis there.

11.3.3 Willingness to Engage in Modern Development Initiatives

Another aspect of human capital that has underpinned the recovery of Gorno-Badakhshan following 1991 is the way in which the population has embraced development. The people there had not just waited passively for outsiders to develop it. Because of this positive approach, reconstruction efforts by AKDN and others have been far more successful than they would otherwise have been. There was a convincing participation of local people in focused infrastructural development efforts facilitated by AKDN, and the distribution of aid. Corruption among the population and officials was observed to be negligible, this having been a necessary condition of both partners for such cooperation. Such "cooperation" was not simply the re-emergence of the collectivization norm under the Soviet era, but by contrast comprised the emergence of the private sector ethic. This latter needed, and still needs, to be channeled into market-led agricultural and livestock enterprise development, based on the new norm of private land ownership. Economies of scale had and have to be visualized and captured to break out of the subsistence-mode individual production on micro-plots, involving value-chain analysis, and adding value at every stage of the chain.[95]

11.3.4 Country-Wide Context

Risk-taking in post-Soviet Tajikistan is still in its infancy, not only as Bliss found in Gorno-Badakhshan, but country-wide. Yet profitability of low-value food crops for the local market, such as wheat, is eclipsed by many other crops, which vary in their associated risks; the lowest risks are often with crops that are relatively non-perishable, such as onions. Yet perennial fruits are often the most profitable crops per unit of land, hence generating the income necessary to keep a family food-secure through its ability to economically access far more staple foods than it could have by growing them on its limited land holding.

As an example of how old ways of thinking are difficult to shift, and the resistance to take risks, an entrepreneur farmer whom the current author met between Dushanbe and the Uzbekistan border in 2015 was reluctant to install a generator back-up for his refrigerated cold stores, to cater for the "load shedding" from the national grid when river levels and associated hydropower generation reduced during the annual dry season.[96] He was adamant that the public sector would soon provide permanent power, by installing a priority "red line" supply to his farm. He had been waiting a while for this, yet dreamed of it happening soon—a cut-away to old Soviet

---

[95]This was observed to be happening in the lower plains of Tajikistan when the current author worked there in 2015, with inventive ways employed by farmers and traders to bypass some of the command economy features remaining in the public sector which stand in the way of agribusiness flourishing.

[96]http://www.bbc.co.uk/news/world-asia-37929367 (accessed December 15, 2015).

times when government did provide well for its satellite State citizens, albeit not in an economically-sustainable way.

One very successful farmer, who had installed at his own expense both refrigeration and back-up generators, explained to the current author that as a result he was able to delay marketing, with a 1-month delay after peak table grape harvest leading to a trebling of his income. In late 2015, while visiting the main fruit and vegetable market in Osh (on the old silk road in the Fergana Valley of neighboring Kyrgyzstan), the author witnessed that by far the best grapes on offer were beautifully-packaged and clearly labeled "produce of Tajikistan". There are excellent export opportunities for grapes and other temperate high-value fruits too in other regional markets, such as in Russia and Kazakhstan.

Most of the fruit "farmers" in the Dushanbe plains are first-generation farmers, largely retrenched public sector servants who needed to create self-employed opportunities in agriculture as the best livelihood strategy to support their families, on the small plots of land allocated them by government. They benefit from hiring space in cold storage facilities that larger farmers have and using any residual space on private sector haulage trucks that would come to larger farms for their "outgrower" produce. Grape production by these outgrowers is too small to entice private hire export business haulage trucks; only by pooling their (high-quality) produce with that of the large farmer can they market their grapes in a timely way, so important for such a perishable commodity. Economies of scale are thus achieved.

Another aspect of such was cost saving through bulk buying of packaging material and genuine agrochemicals; the latter is so important, for many of the agrochemicals which are imported into Tajikistan through official channels are not genuine—diluted active ingredient, completely bogus, and/or even harmful to the crop. Many of the labels on agrochemical packaging in agricultural supply shops in Tajikistan are in Chinese rather than Cyrillic (Russian) script; the former is unintelligible to most educated first generation farmers in Tajikistan. It is estimated that even in mainland Europe some 30% of agrochemicals are adulterated, so it is unsurprising that in Tajikistan the majority is thus classified. Sourcing genuine chemicals at affordable prices often involves a middleman from the exporting country (Kazakhstan or Russia, say) visiting farmers in Tajikistan to build their confidence, taking their orders and part payment, returning to his home country, and then accompanying the imported consignment over the border, doing the necessary with the customs officials, so that the prohibitively high import duties are waived or reduced. Such are some of the horrors of engaging in agribusiness in Tajikistan.

A major export from Tajikistan to Russia has been its skilled labor, yet because of "Western-imposed" sanctions on Russia in recent years, many of this Russian-speaking labor force has been retrenched from Russia. This has resulted in the curtailment of a vital source of foreign exchange through expatriate remittances, for both the national coffers and family support, while simultaneously creating incremental human capital resources for local income generation at home.

REFERENCES

Agarwal, B., 1994. A Field of One's Own: Gender and Land Rights in South Asia. Cambridge University Press, Cambridge.

Ali, M., Byerlee, D., 2002. Productivity growth and resource degradation in Pakistan's Punjab: a decomposition analysis. Econ. Dev. Cult. Chang. 50 (4), 839–863.

Allen, P., 1999. Reweaving the food security net: mediating entitlement and entrepreneurship. Agric. Hum. Values 16, 117–129.

Australian National Audit Office, 2014. Food Security in Remote Indigenous Communities. Department of the Prime Minister and Cabinet, Canberra. September 25, 96 pp.

Bailey, R., Wellesley, L., 2017. Chokepoints and vulnerabilities in global food trade. Report by Chatham House, London. June, 124 pp.

Baumert, K.A., Herzog, T., Pershing, J., 2005. Navigating The Numbers: Greenhouse Gas Data and International Climate Policy. World Resource Institute, Washington DC.

Berry, J.W., 1997. Immigration, acculturation and adaptation. Appl. Psychol. Int. Rev. 46 (1), 5–68.

Blaikie, P., Brookfield, H., 1987. Land Degradation and Society. Methuen, London.

Bliss, F., 2006. Social and Economic Change in the Pamirs (Gorno-Badakhshan, Tajikistan). Routledge, Oxford. 378 pp.

Breusers, M., 2001. Searching for livelihood security: land and mobility in Burkina Faso. J. Dev. Stud. 37 (4), 49–80.

Brody, H., 2000. The Other Side of Eden: Hunter-Gatherers, Farmers and the Shaping of the World. Faber and Faber, London. 400 pp.

CCC, 2015. Progress in Preparing for Climate Change: 2015 Report to Parliament. Committee on Climate Change. June, 244 pp https://www.theccc.org.uk/wp-content/uploads/2015/06/6.736_CCC_ASC_Adaptation-Progress-Report_2015_FINAL_WEB_070715_RFS.pdf. Accessed January 9, 2017.

Cleveland, D.A., 2014. Balancing on a Planet: The Future of Food and Agriculture. University of California Press, Berkeley, CA; London. 320 pp.

Clower, R.W., Dalton, G., Harwitz, M., Walters, A.A., 1966. Growth Without Development: An Economic Survey of Liberia. Northwestern University Press, Evanston, IL.

Conference Board of Canada, 2017. Canada's Food Report Card 2016: Provincial Performance. . Ottawa, May 18, 157 pp.

Conway, G., 1999. The Doubly Green Revolution: Food for All in the 21st Century. Cornell University Press, Ithaca, NY.

Croal, P., Darou, W., 2002. Canadian First Nations' experiences with international development. In: Sillitoe, P., Bicker, A., Pottier, J. (Eds.), Participating in Development: Approaches to Indigenous Knowledge. Routledge, London, pp. 82–107 (Chapter 5).

Dungait, J.A.J., Cardenas, L.M., Blackwell, M.S.A., et al., 2012. Advances in the understanding of nutrient dynamics and management in UK agriculture. Sci. Total Environ. 434, 39–50.

Edmondson, J.L., Leake, J.R., Davies, Z.G., Gaston, K.J., 2014. Urban cultivation in allotments maintains soil qualities adversely affected by conventional agriculture. J. Appl. Ecol. 51, 880–889.

Fedoroff, N.V., et al., 2010. Radically rethinking agriculture for the 21st century. Science 327 (5967), 833–834.

GACGC, 1994. World in transition: the threat to soils. Annual Report, German Advisory Council on Global Change, Bonn.

Gadsby, P., 2004. The Inuit paradox. Discov. Mag. October 1, 2004. http://discovermagazine.com/2004/oct/inuit-paradox.

Goodman, D., 2004. Rural Europe redux? Reflections on alternative food networks and paradigm change. Sociol. Rural. 44 (1), 3–16.

Hopkins, R., 2008. The Transition Handbook: From Oil Dependency to Local Resilience. Green Books, London.

IAASTD, 2009. Synthesis Report With an Executive Summary: A Synthesis of the Global and Sub-Global IAASTD Reports. International Assessment of Agricultural Knowledge, Science and Technology for Development. Island Press, Washington DC. 106 pp. http://apps.unep.org/redirect.php?file=/publications/pmtdocuments/-Agriculture%20at%20a%20crossroads%20-%20Synthesis%20report-2009Agriculture_at_Crossroads_Synthesis_Report.pdf.

INAC, 2015. Indian and Northern Affairs Canada. http://www.inac.gc.ca.

IPCC, 2014. Climate change 2014: synthesis report, Contribution of Working Groups I, II & III to the Fifth Assessment Report of the Intergovernmental Panel on Climate Change, 169 pp.

Kishigami, N., 2004. A new typology of food-sharing practices among hunter-gatherers, with a special focus on Inuit examples. J. Anthropol. Res. 60 (3), 341–358.

Klein, J.A., Pottier, J., West, H.G., 2012. New directions in the anthropology of food. In: Fardon, R., et al. (Eds.), The SAGE Handbook of Social Anthropology. vol. 2. SAGE, in Cooperation with the Association of Social Anthropologists (ASA) of the UK, Los Angeles, CA, pp. 299–312. July 25 (Chapter 4.2.3).

Lal, R., 2004. Soil carbon sequestration impacts on global climate change and food security. Science 304, 1623–1627.

Lal, R., 2010. Beyond Copenhagen: mitigating climate change and achieving food security through soil carbon sequestration. Food Sec. 2, 169–177.

Lee, R.B., DeVore, I. (Eds.), 1968. Man the Hunter. Aldine, Chicago, IL.

Lymbery, P., 2017. Dead Zone: Where the Wild Things Were. Bloomsbury, London. 384 pp.

Marsden, T.K., Banks, J., Bristow, G., 2002. The social management of rural nature: understanding agrarian-based rural development. Environ. Plan. A. 34 (5), 809–825.

McIntosh, W.A., 2013. The sociology of food. In: Albala, K. (Ed.), Routledge International Handbook of Food Studies (2012). Routledge, London & New York, p. 20 (Chapter 2).

McKeon, N., 2015. Food Security Governance: Empowering Communities, Regulating Corporations. Routledge, London. 246 pp.

McMichael, P., 2015. Book review of McKeon, N., 2015. Food security governance: empowering communities, regulating corporations. J. Peasant Stud. 42 (5), 1051–1052.

Melchett, P., 2016. UK Parliament, 2016. Soil Health. First Report of the Session 2016–17. HC 180, June 2, 49 pp https://publications.parliament.uk/pa/cm201617/cmselect/cmenvaud/180/18002.htm (Contains Parliamentary information licensed under the Open Parliament Licence v3.0., October 16, 2017).

Millennium Ecosystem Assessment, 2005. Ecosystems and Human Well-being: Current States and Trends. vol. 1 Island Press, Washington, DC.

Morris, C., Kirwan, J., 2012. Vegetarians: uninvited, uncomfortable or special guests at the table of the alternative food economy. In: Williams-Forson, P., Counihan, C. (Eds.), Taking Food Public: Redefining Foodways in a Changing World. Routledge, New York and London, pp. 542–560 (Chapter 40).

Oldeman, L.R., 1994. The global extent of soil degradation. In: Greenland, D.J., Szabolcs, I. (Eds.), Soil Resilience and Sustainable Land Use. CABI, Wallingford, pp. 99–118.

Oldeman, L.R., Hakkeling, R.T.A., Sombroek, W.G., 1991. World map of the status of human-induced soil degradation: an explanatory note. International Soil Reference and Information Center (Wageningen) and UNEP, Nairobi.

Rickson, R.J., Deeks, L.K., Graves, A., Harris, J.A.H., Kibblewhite, M.G., Sakrabani, R., 2015. Input constraints to food production: the impact of soil degradation. Food Sec. 7 (2), 351–364.

Robinson, D.A., Hockley, N., Cooper, D.M., et al., 2013. Natural capital and ecosystem services, developing an appropriate soils framework as a basis for valuation. Soil Biol. Biochem. 57, 1023–1033.

Rodon, T., Schott, S., 2014. Towards a sustainable future for Nunavik. Polar Rec. 50 (3), 260–276.

Sahlins, M., 1999. What is anthropological enlightenment? Some lessons of the twentieth century. Annu. Rev. Anthropol. 28, i–xxiii.

Schurman, R., Munro, R., 2009. Targeting capital: a cultural economy approach to understanding the efficacy of two anti-genetic engineering movements. Am. J. Sociol. 115, 155–202.

Scialabba, N., Hattam, C., 2002. Organic Agriculture, Environment and Food Security. Environment and Natural Resources, Series 4, FAO, Rome.

Sen, A.K., 1981. Poverty and Famines: An Essay on Entitlement and Deprivation. Clarendon Press, Oxford. 257 pp.

Stefansson, V., 1935. Adventures in diet I, II & III Harper's Monthly Magazine. . November 1935, December 1935, January 1936.

Sumelius, J., Vesala, K.M. (Eds.), 2005. Approaches to social sustainability in alternative food systems. Ekologist Lantbruk NR 47, Uppsala. December, 32 pp.

Surette, M.E., 2008. The science behind dietary omega-3 fatty acids. Can. Med. Assoc. J. 178 (2), 177–180.

Thrupp, L.A., 1998. Cultivating Diversity: Agrobiodiversity and Food Security. World Resource Inst, Washington, DC.

UK Parliament, 2016. Soil Health. First Report of the Session 2016–17. HC 180. June 2, 49 pp. http://www.publications.parliament.uk/pa/cm201617/cmselect/cmenvaud/180/180.pdf. Accessed January 9, 2017. (Contains Parliamentary information licensed under the Open Parliament Licence v3.0., http://www.parliament.uk/site-information/copyright-parliament/open-parliament-licence, October 16, 2017.)

UNEP, 2007. Agriculture and the environment. In: UNEP, Sudan: Post-Conflict Environmental Assessment. United Nations Environmental Program, Nairobi, pp. 158–191 (Chapter 8).

Wenzel, G.W., 2013. Inuit and modern hunter-gatherer subsistence. Études/Inuit/Studies 37 (2), 181–200. https://www.erudit.org/revue/etudinuit/2013/v37/n2/1025716ar.html.

WMO, 2017. WMO Statement on the State of the Global Climate in 2016. WMO-No 1189, World Meteorological Organization. 28 pp.

Wood, S., Sebastian, K., Scherr, S.J., 2000. Pilot Analysis of Global Ecosystems: Agroecosystems. IFPRI and World Resource Institute, Washington, DC.

World Bank, 2004. Agriculture Investment Sourcebook. World Bank, Washington, DC.

CHAPTER 12

Conclusions

Contents

12.1	One Person Can Make a Difference	211
12.2	Social Component Essential	211
12.3	Participation	212
12.4	External Actors in Development Design and Implementation	213
12.5	A Role In-Waiting for Social Anthropologists and Their Ilk	214
12.6	Building the Team Toward a Common Narrative	215
12.7	Sustainable Food System	216
12.8	Two Other Generic Priorities	217
	Reference	218

12.1 ONE PERSON CAN MAKE A DIFFERENCE

The title of this book is *Human Resilience against Food Insecurity*, a follow-up to the earlier book *Food Security in the Developing World*. The author has endeavored to show that along the development road are many choice points, and each of these can be influenced by every one of us involved. One person can make a difference; a good team can make an even bigger difference. "All that is necessary for the triumph of evil is that good men do nothing", said Edmund Burke, Irish philosopher and politician. Food insecurity is one such evil. Human resilience against it is bolstered by wise decisions, impeded by poor ones, and the book considers some of each category, regarding how we make them, and how we might improve them (see Foreword).

12.2 SOCIAL COMPONENT ESSENTIAL

There must be a social component to internationally-supported development initiatives aimed at achieving food security, for these should not merely address technical issues. There is more than enough food produced for us all in the world, but it is not distributed to all the places where it is needed. A major reason for hunger and undernutrition is not global scarcity

Human Resilience Against Food Insecurity
https://doi.org/10.1016/B978-0-12-811052-2.00012-3

of food, but scarcity of social justice and global democracy, leading to widespread poverty and hunger. And not only in the developing world—there is increasing inequality in developed countries too.

At the local level, change in behavior is an indicator whereby permanent change may be measured; this is needed as a determinant of resilience. This will not happen by lecture, imperative, training or even demonstration alone. There has to be a change in mindset, in motivation of the "beneficiaries". For this to happen they need to have ownership, and for that to happen the implementing team needs to understand the point at which the target community is now.

The project must have full empathy with the beneficiary party, to become that party, to attain Rilke's "in-panther" feeling so he or she can think as a panther does. To get onto the same part of the planet and thereafter the same wavelength, is a process, and this takes time, there being many a potential tumble and fall along the way, as teased out in Chapters 4 and 5. The process requires energy and dedication. I am told by a member of the social sciences team in the Cambridge University Global Food Security Initiative that he and his colleagues hear from their plant scientist colleagues that "social scientists have a problem for every solution!". Yet the new tools that plant scientists create are embedded in a social context. The two perspectives do need to hold a conversation and negotiate.

12.3 PARTICIPATION

For such negotiation to happen, from the project identification kick-off point in the project cycle, local beneficiaries need to be consulted. This means not simply their being told what is going to happen to them, but for them to be asked what it is they need, what their priorities are, and which of these are most suitable for a project to deliver over a given time frame. The beneficiaries need to participate in identifying the objectives and actions, how achievement of these may be measured, what the beneficiary community is willing to contribute, and what help it needs from outside to deliver against the objectives and outputs. Their expectations of the project need to be tempered with reality, in terms of scope, quantity, and timeliness, and the potential constraints that can be anticipated and dealt with understood, as well as those potential constraints that can only be guesstimated, and that some will be outside of the scope of the project to address.

Participation in such planning and subsequent implementation should serve as a tool for the empowerment of the beneficiary group, so that through its involvement it secures ownership of the benefits and outcome of the intervention. Moreover, active participation engenders confidence in communities that they can upscale these benefits in perpetuity, and on their own devise other start-ups from which they may prosper. True participation should comprise a key component of a donor's exit strategy. Such holistic participation is woefully uncommon in projects, as Bliss and Neumann have pointed out (Box 6.2); planners within donor and development agencies in the developed world are largely to blame for this.

12.4 EXTERNAL ACTORS IN DEVELOPMENT DESIGN AND IMPLEMENTATION

External actors need to be nurtured in the norms of the community with which they will be working. They must be prepared to learn and adapt their home-base thinking and behavior if they are to gel as seamlessly as possible with the local community. This may be more challenging for those in the management category who are older and set in their ways, and for those who have been entrenched in an institutional culture to such an extent that their ability to be creative and adaptable has been seriously compromised.

Chapters 4 and 5 in this book encourage expatriate development professionals in the field to find out more about the nationals with whom they are working, their partners in development, their mindsets and livelihoods, before they jump in trying to impose their "solutions", which may well be inappropriate supply-led rather than demand-led interventions. Let us first find out why the target community thinks as it does and does as it does, to understand better its diplomatic baggage. We should not assume that the reference frame is the same as one would expect in a community in developed countries, for after all, reference frames vary widely there too, within-country and between countries. Let us all be imaginative, as advocated by Joanne Rowling in her Harvard presentation in 2008 (Section 10.4), and merge the best of what is new with the best of what is old, and vice versa (Section 8.9). Last but not least, let us listen to the voice of poetry, and endeavor to retain the "sacred balance" to which the late Darrell Posey refers (Section 8.1), and reflect on "augmenting methodology with magic" as advocated by poet Dana Gioia (Chapter 1). In so thinking and doing, the resilience of both the environment and the social fabric is maintained and fostered in the pursuit of food and nutrition security.

Those who design food security projects need to get these right, identifying both the priority choke points, and how best to resolve them. Getting it right has not always been the case (Section 6.2.2), one reason being that it is "foreigners" who often do the designing. Listening to and learning from well-informed and elected opinion-leaders in the target country or region should be obligatory.

Downstream of design, during implementation of these projects a rational, independent, and empathetic approach is needed from outside agencies in their dealings with citizens of developing countries (and transition countries, and marginalized groups in developed countries). This applies most particularly to the formidable institutions that are governing developing countries from afar, especially transnational financial institutions and development agencies (UN agencies and donor organizations) (Section 6.2.1). These powerful largely unaccountable transnational institutions have been little studied, yet play a central role in the *de facto* governance and administration of their partner countries. The way that their policies and priorities are determined may legitimately be called into question, as indeed the nature of their unequal relationship with those countries, which can be overly didactic and reflect more the interests of their powerful financial backers than the interests of people in the developing countries concerned. The oft top-down disposition of these agencies has been characterized by some anthropologists and political scientists as "foreign hegemony" (Section 8.1). These agencies need to get it right more often than is currently the case.

12.5 A ROLE IN-WAITING FOR SOCIAL ANTHROPOLOGISTS AND THEIR ILK

At the boffin level, social, cultural, and organizational anthropologists could and should be instrumental in mentoring their colleagues in matters social and cultural—in the biological and economic sciences, for instance, and both technical and non-technical administrators in development banks and donor agencies. Yet first, the social anthropologists themselves may need to rethink their own methods of understanding and cataloguing. In the preface to the book by Rose (1990), the three distinguished editors call for a new type of research and reporting style across the social sciences, not just in anthropology. They count themselves with critics who argue that: "conventional social research dulls the imagination, locks the observed inside rigid category systems having little to do with the culture of the researched, but everything to do with our research culture; promotes an insidious institutionalization of social boundaries that separate 'us' (the observers) from them (the observed); and perhaps most telling has become rather tedious, if not boring, thus losing its power to convince". In short, past practices have "become socially and intellectually suspect". We need "ethnographies of intimacy not distance, of stories not models, of possibilities not stabilities, of contingent understandings not detachable conclusions". What we need is "genre-busting ethnography".

Rose himself argues that the postmodern moment has arrived in social sciences. He maintains that ethnography is overly linked with corporate purposes (academic and economic). He suggests that pretextual assumptions passed from one generation to the next more or less determine what it is we investigate and therefore what we say about those ethnographic worlds. From the Western side and contemporary ethnography, anthropologists need to elicit but not dominate the ideas and practices of the people that Westerners study.

Addressing anthropologists in particular Rose says that "ethnography as an entity grew up in the colonizing praxis of capitalist culture" (his page 10). He joins those "who in growing numbers question the assumptions and practices of anthropology against the necessity to decolonize academic thought" (page 15). "Doing ethnography is to construct a text from the experiences of others, experiences carefully controlled by the profession", he says on his page 14. In the work he did among the black communities in the United States, he couldn't gather data in the traditional mode of his profession, as he lived with those he was studying there, as one of them. He had to unlearn the scientific method he had been taught at college, which latter involved applying the received assumptions he learned there (his page 11).

It would be beneficial to have more female social anthropologists in the field too, to avoid the potential gender bias on types of information collected. For instance, Gadsby *ibid.* (pers. com.) opines that the importance to nutritional resilience of seasonal fruit and wild plant collection by traditional Inuit women has been underrepresented in the literature, because (the almost exclusively male) explorers and anthropologists have tended to "hang out" selectively with their menfolk and join in their hunting forays.

12.6 BUILDING THE TEAM TOWARD A COMMON NARRATIVE

To address the resilience priorities requires application of "joined-up-ness", across all stakeholder groups, this requiring more than just scientists and anthropologists working more closely together as advocated in Section 12.5. The Common Narrative initiative in Bangladesh was cited in Section 8.1 as a good example of that, as also industry-wide efforts to enable remuneration for tea pickers to reach living wage levels (Chapter 1). A third example is that in 2017 for the first time, UNICEF and the WHO have joined the FAO, IFAD, and WFP in preparing the annual State of Food Security and Nutrition in the World report. This reflects the broader view on hunger and all forms of malnutrition promoted by the agenda of the Sustainable Development Goals.

At planning and implementation level, the importance of local knowledge and experimentation, and applying a degree of lateral thinking is advocated in Chapters 8 and 9, with a range of examples given. The role of local champions (Chapter 10) is hugely important, often making the difference between success and failure, between post-project non-sustainability and a surge in community social and human capital so crucial for long-term resilience against the scourge of hunger and undernutrition. Moreover, as indicated in Section 12.4, the powerful international development institutions holding the money bags on behalf of the international community need to ensure they always listen to, and plan with, national stakeholders.

Once we have got our act together as development workers, and have devised an enhanced "joined-up-ness", we are better positioned as a team to conceptualize, identify, and prioritize what should be the best foci for our efforts to improve resilience against food insecurity (Chapter 6).[97]

12.7 SUSTAINABLE FOOD SYSTEM

Sections 6.1 and 11.1 urge a more sustainable food system(s) to be better identified and practiced in both developed and developing countries, in part through better recycling of nutrients and integration of crop and livestock enterprises, to maintain soil fertility and enable food availability needs to be met. On-farm operations cannot remain on a business-as-usual basis; overwhelming scientific evidence dictates that.

A second aspect of current food system unsustainability refers to the many people in the world who are food-insecure through no fault of their own, and despite their best efforts. Marginalized people like my guest-house cleaner in The Gambia (Section 4.5.9) should not have to choose whether to channel their disposable income into shelter *or* food, for they may be unable to afford both. This inequity has resulted from an unfair distribution of the world's bounty, and one does not need to be a Marxist to accept that. This is the depraved "circumstance" which Sridhar highlights (Section 6.2.2).

---

[97]In addition to gender transformation, which was accorded top priority in the author's earlier book.

Two specific aspects of seeking a sustainable food system could exercise us more than has been the case thus far. First, anthropologists and sociologists should review what work has been done on the psychological and ethical constructs of the affluent society, and delve into it more. There is a massive amount of research already done on its antithesis, poverty and deprivation—how much more is there to learn about the fundamental predispositions to, and horrors of, these? What about the affluent society in this consumer age, where what we have, own, and throw away (food included) seems to define who and what we are, and advertising hype demands we must procure the very latest fashion accessory? Fortunately, we do not all fall into that trap, even if we can afford to do so (Section 6.1), but why do so many do so? The "nudge theory" of Thaler seems to offer great promise in helping us all be less wasteful and more considerate of the global future.

Second, archaeologists should be brought more into the mainstream of food security research and practice. At the start of this book is a quotation from Shakespeare's *The Tempest*, in which Prospero exhorts Miranda to delve beneath the surface, to seek the reason for their current condition:

What seest thou else

In the dark backward and abysm of time?

Archeologists and anthropologists are together looking at what has changed at sites of long-term human occupation, in terms of food grown and eaten, and the governance systems which can be divined as pertaining, to help us better comprehend resilience or otherwise, then and now (Section 8.9.1). Ways to maintain resilience were there in the "dark backward and abysm of time", some since abandoned, and there is likely much that we can and need to learn from that Rosetta Stone of knowledge and practice. Likewise, some of the governance and agricultural extension systems we used to follow in more recent times did well for us, with soil fertility and farming systems being sustainable. We need to look again for, and at, those systems.

12.8 TWO OTHER GENERIC PRIORITIES

In addition to a more equitable food system, better policies are cited as a priority area for closer attention: those which better help facilitate project design and implementation—from governments, development banks and donors—and enable beneficial outcomes from these to be sustained and upscaled (Section 6.2). Examples are given where policies and planning are good, and when they are bad.

Finally, another priority cited is for more focus on engineering, then capturing, the peace dividend (Section 6.4). This involves more intelligent peace-making initiatives.[98] The approach has already been espoused by the "sociology school" of West Point military college in the United States, which (through General Petraeus and others) rendered more intelligent the way that the International Security Assistance Force (ISAF) military campaign was waged in Afghanistan. Surely it is worth putting at least as many resources into intelligent peace-making as into intelligent warmongering. To build resilience against food insecurity and undernutrition in conflict-affected situations requires a conflict-sensitive approach that aligns actions relating to humanitarian assistance, long-term development and sustaining peace.

Poets have excelled at expressing the feelings of the dispossessed and disenfranchised, though perhaps they are less good at applying their art and sensitivities to proposing solutions and steering them through. This latter role seems to me to be an area in which they could excel, working hand in hand with social scientists across the board toward social and political reform, applying a soothing balm to help subdue pain and heal deep-seated wounds, thereby facilitating an enabling environment in which negotiation can flourish. There can be no long-term resilience against food insecurity in the presence of conflict, as acknowledged by the UN Report "The State of Food Security and Nutrition in the World 2017" *op. cit.*

There is work to do in plenty in making and keeping the peace, by anthropologists, sociologists and poets, and even foresters and plant pathologists (Section 8.1), who reach out into what may at first seem unfamiliar social science territory, in response to Prospero's challenge: "What seest thou else?".

REFERENCE

Rose, D., 1990. Living the Ethnographic Life. Qualitative Research Methods Series, vol. 23 Sage, Newbury Park, CA. 64 pp., pp. 5–6.

---

[98] I came across an example of this in the United Kingdom. An anti-Islam group had gathered outside of a mosque in York, England in 2013. It was defused by the brilliant expedient of some of the Muslims inside coming out and saying something like "You must be tired. Would you like to come in for a cup of tea?". And they did.

ANNEX 1

Goats and Nightclubs of the Levant

Despite being a seasoned man of the road, I reflect that it was only when I worked in southern Yemen that I had to push two reclining goats off the dining table before I could be served my meal. I was in the town of Shabwah in the Hadramaut region, reached by a winding ribbon of road across the barren craggy escarpment rising abruptly from the Gulf of Aden to the plateau above. The restaurant, which would hardly feature in a connoisseur's guidebook, was a typical sort of al fresco restaurant in a typical sort of Arab outback town, with petty service development along each side of a dirt street. Its surface was compressed by centuries of pounding by human feet in search of rivets, tobacco, frankincense or soap, and the eccentric explorer Freya Stark herself in the 1930s in search of adventure and discovery, and maybe another lover.

And of course the road was pounded also by goats, sheep and camels, and at night by bands of scavenging jackals and perhaps the occasional aging Arabian leopard, on the lookout for an instant meal in the form of a tethered goat. The aroma of the street was equally typical, and when I make mention of it to acquaintances in the developed world, it normally proves a conversation-stopper, something which startled me in the early days. How was it possible for anybody *not* to thrill at the mention of a scented blend of roast coffee beans, spices and goat urine? Ever since 1976, when I attended the exhibition called "Sana'a" at the Museum of Mankind,[98] part of London's Year of Islam, and being seduced therein by a simulated street bazaar in northern Yemen, the associations of that heady aroma are for me a source of endless wonder.

For instance, it was in south Yemen during my first visit there in early 1986 that my camera broke. This was just after the month-long South Yemen Civil War that started in January, and the Christmas decorations in my hotel on Aden's Slave Island had not been taken down, such was the trauma of the population. I asked around and was pointed in the direction of a particular

[98] I still have and treasure the brochure of that event.

shop in a dusty side street. It was highly inauspicious, I thought as I entered, that there was not a camera in sight in either the window display or inside. Some specialist shop, this! A tall gentleman of composure appeared from the world behind the counter, and I explained that the shutter was jammed. I was advised to come back the following day to collect the camera. I thanked him, yet without hope that when collected the fault would have been fixed. The next day I duly did as bid and was handed the camera, in perfect working order. I was aghast and ashamed. I thanked the gentleman profusely and asked how much I owed him. "From which country do you come?" he enquired in perfect English. "I was born in Britain", I replied. "Then you don't need to pay anything, do you?"

Many times I've told that story down the years, each time my voice breaking I think, out of love for southern Yemen and its people. During the days of the British colony of Aden and the protectorate of the hinterland around Aden (1937–63) (the area later re-branded as the independent People's Democratic Republic of Yemen (PDRY), which lasted from 1970 to 1990), a bond was welded in the dry heat between Brits and Yemenis, not just of respect as between Brits and Zulus in South Africa, but of reciprocal love too. I was frequently told by southern Yemenis that they felt themselves British. I have no doubt that the camera repair man also kept goats, though I did not ask.

On that first trip to Yemen my government counterpart Abdulkarim and I became dear friends. He featured in a photo I took that was included in a book I subsequently wrote on dryland farming, standing amidst a caked and peeling wadi flow mud residue. I did not meet him again until 25 years later, in Aden, as my trips to Yemen in the meantime were to the north (the Yemen Arab Republic, or YAR, as it was from 1962 to 1990, before being forcibly "united" with the south, as the Republic of Yemen). Another photo in my book showed Yemeni goats grazing on the stubble of the delicate miniature cereal crop called *teff*, which can grow and mature on a single rainfall. I know that Abdulkarim and his family keep goats.

Goats played a vital role in another British colonial era theater in the Levant too: Cyprus. The former UN Undersecretary-General Brian Urquhart explains that when the natives became particularly restless under the colonial yoke, the resident British military would parade its regimental goat to calm the situation. Goats were very much part of the livelihood system in Cyprus at the time, and the sight of one particularly splendid shaggy specimen being paraded, bedecked with ribbons and regalia, adopted and elevated by the colonial administration to near mythical status, accorded recognition of its high esteem in the eyes of both Cypriots and Brits, a co-alescing of cultures.

Further back in history, the multitasking Pan (God of shepherds, flocks, nature, hunting, theatrical criticism and music) is commonly portrayed having the torso and arms of a man, yet the legs, horns and ears of a goat. He pranced through the mountains and countryside frolicking with nymphs at every opportunity while playing his seven-reed pipe, the Pan flute or syrinx. His demeanor and piping were usually soft and seductive but, when angered, his awesome bellow could be heard for miles. His name is still associated with fear (PANic) because of this. By day he hunted and killed predators that menaced the flocks, retreating to his mountain bower by night, where he played the syrinx, its tune more captivating than those uttered by the fowls of the air, and to which the nymphs danced and sang. To avoid the allegedly unwanted advances of the lusty Pan, the nymphs were commuted into various forms to escape him. Methinks that therein lies a sub-myth, that nymphs are never wont to be lusty themselves.

My intrigue with the Pre-Raphaelite school of painting started during my undergraduate days. No painting means more to me than Holman Hunt's *The Light of the World*. Another of his masterpieces is *The Scapegoat*, the black beast which compels the viewer, through the agony revealed in its bloodshot eyes, to empathize with its pained persona within, as it languishes in the slurry at the edge of the highly saline Dead Sea, unable to drink therefrom and in its last moments of life. Through an interesting turn of events when I lived in Jerusalem, and a telecon with renowned Jerusalem newspaper correspondent Eric Silver shortly before he died in 2008, I made a friend of Mira who lives in the house next to the Silvers. I was invited into that house, built and lived in by Hunt, and in which *The Scapegoat* was painted. What an experience for me. A little later in England, at an exhibition in Manchester's art gallery I was able to see *The Scapegoat* together with the three versions of *The Light of the World*, to examine them all, brushstroke by brushstroke. That day was and always will be priceless for me. Eyeballed by *The Scapegoat*, one can internalize Rainer Maria Rilke's "in-dog" or "in-panther" feeling—in this case, "in-goat" feeling—to actually *become* that goat in its hopeless situation, rather than being merely someone confronted by a painting of a goat.

As many readers will know, the image in *The Scapegoat* is based on the Jewish custom of releasing a goat into the wild on what I understand to be the most solemn and important holy day of the Jewish calendar, Yom Kippur or Day of Atonement. Many of those goats would presumably have come to a sticky end in the dry wilderness, as depicted in the painting. In the Old Testament, on that day the High Priest makes a sacrifice to God for the sins of the people, bringing reconciliation between the two, indeed this being a harbinger of the Christian Easter story. Following the blood sacrifice, a goat was released to symbolically carry away the sins of the people. This "scapegoat" was never to return, and the term is still in common usage in Western society.

Maybe this then is the key to the mystique of goats of the Levant, as per the title of this essay. The goat-man God Pan of the ancient Greeks, the scapegoat of Judaism, parading the regimental goat in colonial Cyprus, and so on. And the importance that each story has in the fabric of society comprises part of the cultural glue which binds a people, a nation, together.

But what of the mystique of Levantine nightclubs? You might by now be wondering when I will say something on them, and the link with goats? You may believe I am a latter-day Alistair Cooke, who often left listeners to his *Letter from America* on BBC Overseas Service and British local radio, wondering when, if ever, he would get to the point.

Listening to the Lebanese singer Ferouz (real name Nouhad Wadi Haddad), in Beirut when she was young, her acolytes might contend was almost in the category of the Love of God, which passeth all human understanding. She held the command of an audience par excellence, capturing its corporate breath and holding it in trust for ever, a stunning presence caught in a silver spotlight as she flowed along the catwalk through a spellbound auditorium. She has inspired devotion and adulation throughout the Arab world, then and now. Her iconic technical virtuosity is legendary; even now in her 70s when she still gives the occasional performance, in Lebanon or elsewhere. Let the new listener from the West first sample her repertoire perhaps with *Li Beirut*, a haunting adaptation of the guitar composition *Concierto de Aranjuez* by Joaquin Rodrigo, a moving lament about war having extinguished the lamp of her beloved once-beautiful city, while the youngsters can dance to her *Oudak Rannan*.

Furthermore, I could share with you an evening I spent at the Greek Club in Cairo discussing the work of Egyptian Nobel Laureate in Literature Naguib Mafouz with his former assistant, together with a lady Reuters correspondent....

But back to goats for a while longer... It is probably safe to say that for most people in the developed world, the subject of goats is not one to be tabled in polite society. Yet in most of the developing world, goats are a major pillar of the livelihood system, enabling an instant revenue stream at short notice, representing a whole banking system on the hoof in often-remote places. They illustrate a difference in perspective between two contrasting macrocultures. The bad name attracted by goats for denuding the vegetation of the drylands is misplaced, for goats are often taken to browse after cattle have consumed all that they can find. What is the poor goat to do but eat what is left? If those rangelands are overstocked, as they commonly are, that is not the goat's fault either.

The two civil wars in Liberia between 1989 and 2003 all but wiped out the entire goat population. When I worked there on a USAID-funded Economic Corridors project in 2011, on a field trip from Monrovia to Grand Bassa county in the south we passed only two goats in a whole day's driving, quite remarkable for sub-Saharan Africa. I was moved to comment to my colleagues in the vehicle at the time that we should put the two of them in touch through Facebook, so they could start breeding. During the conflict, the pre-war goat numbers were decimated as goats were a key source of food for a nation trapped in a dislocated economy, with rampant internal displacement of its citizens. There was also a convergent pressure to eat or sell the animals, ahead of their being stolen as war booty by one of the rampaging militia groups.

As in Africa, goats are a major source of food and cash in much of Asia. A scientific colleague of mine on a rural development project in Nepal in the late 1980s engaged in pleasantries with his neighbor during the inbound flight from Frankfurt, a conversation which started thus: "I'm So-and-So, and I'm going to Nepal to inseminate goats with elite stock semen. And yourself?" To which came the stunning reply: "I'm Reinhold Meissner and I'm going to climb Everest's north face". Meissner is recognized as the greatest ever mountain climber, the first to scale all 14 8,000+ft peaks and the first to ascend Everest without carrying a supplementary oxygen supply. He was born in the South Tyrol, an alpine man through and through, so I am confident that he knows a thing or two about goats and can hold his own on the subject—in contrast to my colleague's inability to reciprocate on high altitude climbing techniques.

For myself, I happen to have a farm in Uganda, which my family has owned for almost 30 years but on which I have only recently taken up residence, to support my farm manager. Because of other duties around the world, until now I've been what in Kenya is called a "telephone farmer", exerting management authority by remote control. Needless to say, it doesn't work that well. Much of my farm has been used as common land by the community for decades, and one time-consuming aspect of my work now is to seek to change the culture of some of my neighbors so they understand that the land actually belongs to someone, and that someone is my wife.

One of the early episodes involved abating the nuisance of a neighbor's tribe of goats which had been caught red-handed and unsupervised in an area I'd just planted with tree seedlings, which they were nibbling. I adopted the strategy of arresting the wily leader of the pack, a magnificent buck. One of my workers was a deft hand in such an exploit, and very cleverly lunged·at the goat when he was on an old termite mound, catching it by a hind leg. There ensued an heroic struggle by him and another worker to subdue the animal; they finally lifted it, two hooves apiece, onto my pick-up truck. We then chased the leaderless tribe out of sight. Possession of the buck gave me considerable bargaining advantage in subsequent negotiation with the owner, in pursuit of a charge of dereliction of care for his neighbor.

Being in close attendance to the aforesaid struggle between men and beast I was struck by the strident scarlet mucosa in the nostrils of the up-ended animal, flared in indignation, not to mention humiliation in public—that public in particular being his harem of a dozen or so does which had also infiltrated my land. The vibrant color of that mucosa reminded me of the mouth of Lawrence Olivier, open wide on the stage of London's Old Vic theater as he played Othello in 1967. That embellished redness was probably contrived using a meat reddener, for theatrical effect. The nuance of such lateral thinking in that Ugandan field brought me such pleasure, taking me back decades to buying cheap tickets with my college classmate Mary Denley which placed us behind a pillar, around either side of which we had to peer from the balcony onto the stage and indeed down the throat of Sir Laurence, as he exploded in hysterical rage to the Gods at his misfortune, when Iago tells Othello that Cassio has admitted to having an affair with Desdemona. Just watching Larry's extravagant curtain call bows was worth the price of the seat many times over.

Independent of the tussle with the buck, my elder son had established, also *in absentia*, a small enterprise of rearing goats on our farmland. Husbandry of these animals was entrusted to an old man called Michaeli. He was overly-fond of the local brew, related to which he sadly departed this life during the early days of my hands-on managership of the farm. The responsibility of providing disciplined TLC for four goats then fell to me. There would have been five on which to lavish TLC had not one been slaughtered to feed the guests at Michaeli's funeral.

I was delighted at the challenge on offer, more because this was an opportunity for me to offer TLC to my son while he was pursuing postgraduate studies at the Sorbonne, rather than any interest I then had in raising goats. On my first day on the job I gave his four goats a feast of which they could hitherto only have dreamed, tethering them on my farmhouse unkempt lawn with leashes just too short to enable them to eat my mango and passion fruit seedlings. I knew that the mature female was pregnant. Being clearly unable to carry both her unborn young *and* the prodigious quantity of roughage she'd shifted during the day, overnight she delivered triplets. It was one of those life-changing moments, for her and me. It gave new meaning to TLC, and I was thrust onto an almost perpendicular learning curve as rookie goatherd. I had been billed as an expert in crops and trees, not goats. I texted my son that his goat empire had almost doubled in size overnight.

For 2 weeks I kept the triplets alive, taking off the stronger males one at a time from the two teats so that the smaller female kid should have her turn, and feeding and watering the dam *ad libitum* to ensure that her milk bar was always open. At the end of that fortnight, however, the female kid turned poorly—she hunched her back, her tail dropped and the frolicsome sideways frisking was supplanted by lethargy. I called the local veterinary assistant who administered an oral dose of an intestinal parasite drug, and a shot of antibiotic in the rump. Within 2 hours there was a transformation in her health. Her tail shot up, and she successfully jostled her two brothers off the mother's teat. I was over the moon, and went off to the field for fencing duties happy indeed. Yet by dusk, she was looking poorly again and had to be carried from the newly-mown lawn to the overnight shed.

In the morning I found her lifeless body in the dust. Later in the day I buried her, in alignment between the farmhouse and the sunset, facing my balcony, between a luxuriant gardenia and an hibiscus. I felt really cut up. Across my mind passed a litany of all the situations of loss in my life. I am finishing this essay on that morning. The mother knows that her daughter's life has gone. She bleats at her loss. She knows that there are now two when before there were three. It is not just the goat kid that has died—I almost said *who* has died. Something in me and my life has died too. I was looking after these kids for my elder son. I let him down. Perhaps there was something else I could have done. I was upset because of "associations". More than 20 years ago I lost what would have been my second child, arriving at the hospital to find a red mess in a stainless steel container outside of the ward. "I saved your wife but could not save your child", the doctor said. We should have gone to the doctor earlier, the day before when there was the first sign of trouble. We didn't. We were too confident of a second normal birth.

I have now identified a suitable person in the village to whom I can divest the goat husbandry responsibility, as it takes more time than I can spare. The challenge I accepted of looking after goats in the interim has certainly had its rewards. I have learned something of their nature, while adding something to my understanding of my own nature, in terms of my capacity for patience in particular.

Fast forward 2 weeks from my baptism by fire, as I cradle the two surviving goat kids to curtail their exploratory meanderings on the 100-yard homeward trudge from being in the field with Mum to their overnight farmhouse quarters, I find myself asking them out loud what they learned at school that day, and they answer by mewing and chewing my shirt with their young teeth. "Now I've told you before boys not to speak with your mouth full…" I offer a goat kid to a distraught child, Nalukwago, my farm manager's 8-year-old daughter. Immediately, her voluminous tears stop, replaced by shy joy and giggles.

After I lost my unborn child I found myself humming "Morning has broken", composed by Eleanor Farjeon to an old Gaelic tune, which I had come to love through Cat Stevens' rendering of it in the 1970s, with a superb piano accompaniment by Rick Wakeman. I don't know why it so repetitively came to my mind, despite my trying to erase it because of its dramatic association. It was a long time before I came to love that tune again, and that was through a visit to the American military cemetery at Madingley near Cambridge.

It was not my first time there; I had taken visiting American friends there before. Yet on that later occasion I went into the chapel and "Morning has broken" was playing. I found it beautiful, in the context of the almost 4000 marked graves in the garden outside, and the list on a long Portland stone wall of American troops missing in action during the Atlantic campaign of the Second World War. My father had worked with the 5th American Air Force in another theater of that war.

I also witnessed something else very moving at one of the graves, which branded itself into my psyche. The incised epitaph on one of the erect gravestones had just been filled with wet sand, clearly from a favorite beach near the family home across the Pond, lovingly conveyed to Madingley by a still-grieving loved one. So now, I once more have an association of that hymn with beauty, through the concepts of love, honor, service to the nation and everlasting life.

Now, what about those nightclubs of the Levant? Their consideration may now seem out of place. I have certainly made reference to the Levant in plenty. Yet here I have to confide in the reader, hoping not to be on the receiving end of an avalanche of hurled shoes, that I had little intention of giving equal weight to both goats and nightclubs, exhilarating though the latter may be. Yet I divined that to expect a readership at all under the laconic title of "goats" was unreasonable. And yet readers *have* read about goats thus far, drawn on perhaps not by the desire to know about goats, but in anticipation of the review of nightclubs which would follow.

With the reader's kind compliance, rather than my delving deeper into Levantine nightclubs, I plan to table for consideration at this point what selection process operates to determine what we read and what we don't, and why. The discourse on goats and matters arising by association was not entirely without merit, was it? There was no technical stuff, no flurry of incomprehensible jargon, no tedious acronyms and certainly no dreaded PowerPoint presentation to send you to sleep. Some readers will have learned something, which may not necessarily have a direct bearing on the main thrust of his or her life, yet which may well have an indirect bearing, if the reader enables that to happen—lets it all hang out, as it were. There is gold in them thar hills by widening the scope of what one chooses to explore, through reading or doing, rather than becoming too focused in order to know yet more about one's own specialty interests, in the expectation of a predictable outcome.

Should we not be rather more experimental, and make time to pursue avenues of experience which might just lead to our finding that gold? The American school education system teaches the value of this, compared with the British system which requires an earlier specialization, with some doors not only closed early but never opened at all. Some of us lucky enough to have had a university education digressed from our coursework to expand our scope of understanding of the world. One of my own triumphs was to take time out from studying science to attend a course of lectures on the poet T.S. Eliot. I am so glad I did. And in *Little Gidding*, Eliot even talks of livestock housing! Those who visit that hamlet know that the pigsty really does exist, as Eliot describes, and indeed compellingly sets the scene for the poem. The wider we search for new knowledge and experiences, the more sure-footed may be our progress and achievements (as sure-footed as a goat one might say) through learning about previously unfamiliar situations, and understanding our response to them and how those changes might be adopted into our aspirations.

So, three cheers for goats, a notional indicator of the richness the world has to offer to those who take the trouble to explore in a somewhat extra-curricular way. Let that exploration not stop once one leaves school. In *Little Gidding*, Eliot offered four lines in particular which have empowered me in a battle-scarred life, and will continue to do so. May I share them with my readers, whom I salute, for most have persevered to the bitter end, reading about a subject in which they may have been fairly sure they would find no resonance:

> We shall not cease from exploration
>
> and the end of all our exploring
>
> will be to arrive where we started
>
> and know the place for the first time.

The *place*, in this case, is no less than the cosmos, the same cosmos as-it-actually-is with which Sylvia Plath increasingly lost contact, as noted by Joyce Carol Oates in her essay "Death Throes of Romanticism". Had Sylvia ventured out more from her sequestered life perhaps she would still be with us, would have better managed her depression and psychotic traits, even though that would likely have meant no final frenzy of autumn and winter poems, or even fame. Yet she might have found contentment.

Goats have provided a theme by which this window of wider understanding can be opened. There is an infinite number of other conduits, and I offer a few others below which have opened *my* eyes. When I lived in a traditional house in the one-street town of Sandikarkha in the western mountains of Nepal, I would be woken long before dawn by the sound of women pummeling maize grain in a mortar, the same women who were last to bed the night before. And why children in mountain villages are often so dirty and their scalp encrusted with infected lice lesions was understood the moment I gasped in shock while taking my first shower in near-freezing water under the garden pump.

In my village in Uganda right now, the sound of metal beating often transcends the whole night. Are these artisans deranged, or just plain inconsiderate for the rest of the community which wishes to sleep? Neither applies. They do this while the national grid power supply is connected... much of the time it does not reach our village, and the artisans cannot ply their wrought iron trade. They have to work when there is access to power to be able to heat, bend and weld, to earn a living by producing what the community needs, especially burglar bars and steel doors for shops and houses, and to meet customer deadlines, no matter that this may mean working through the night.

Edward Thomas, a great Welsh poet, was a close friend of the author of "Morning has broken", Eleanor Farjeon, of whom mention was made above. Thomas was killed at the very start of the First World War in the Battle of Arras, on April 9, 1917. On a loose sheet of paper recovered from his diary he kept in the trenches were two lines which he had perhaps intended incorporating in a future poem. In pencil he had written:

Where any turn may lead to Heaven

Or any corner may hide Hell

Thomas lived both propositions simultaneously in the trenches. They remind us that around any corner there lies in wait some new experience, a morsel of new information, which if acted upon can shape our ends, or those of others, for the better. Bob Geldof and Live Aid are proof of that: the live shows had a global audience of almost 2 billion. He was an expert pop musician, but got involved with Ethiopia in 1984, and became an expert on African famine relief. He jumped the knowledge gap. I've seen him out-debate the then-British Prime Minister Margaret Thatcher on the issue.

Let's respond to Eliot's exhortation to us that there will be no end to our exploring, and through focusing on the unity of past, present and future, all shall be well and all manner of things shall be well. "Only connect" said E.M. Forster in his novel *Howard's End*. I'm on the case. Please join me. Next time you see an article on goats or some other out-of-your-world experience, don't ignore it and pass on to the stock prices, latest gaffe by a presidential hopeful or "entertainment". Education can be entertaining too, more challenging and a whole heap of fun. You never know what you might be missing otherwise. After all, Cinderella was no farmer, yet had she not taken a passing interest in pumpkins she would never have gotten to the Ball...

INDEX

Note: Page numbers followed by *f* indicate figures and *b* indicate boxes.

A

Afghanistan, 60
  absence of peace, 94
  colleagues' wellbeing in, 31
African colonial period, sensitive language
  and customs related to, 33–35
Aga Khan Development Network
  (AKDN), 204–205
Agricultural decision-making process,
  21–23
Agricultural Development Officer
  (ADO), 29, 161
Agricultural Education and Training
  (AET), 91
Agriculture
  at crossroads, 180
  intensification, 183
  livestock production, 174
  monocrop, 51
  production techniques, 74–75
  research
    in developing countries, 142
    station, 158
Agroecosystem, 182–183
Alaska, 192*b*
Anodyne, 1
Anthropologists, 52, 78
Anthropology, 3
Anthropology of Food, 127
Applied archeology, 141
Arctic and subarctic zones, food-getting
  strategies, 50
Arctic Inuit, 131–132
Ashley, John, 102*b*

B

Begging, on streets (Delhi), 5*f*
Beneficiaries, 62, 212
  community, 27, 212
  sensitivities to, 35–37
Bhutan, market scenes in, 114–115*f*
Boko Haram, 116–117

Building resilience
  chronic undernutrition, 16
  food insecurity, 147
  local knowledge in, 125, 127
Building Resilience and Adaptation to
  Climate Extremes and Disasters
  (BRACED), 19–20

C

Call for Proposals (CfP), 119
Cash-for-work schemes, 10
Champion's role
  from commercial private sector, 168–170
  from international organizations, sport
    and entertainment, 171–172
  at village and public sector levels, 167–168
Change management
  institutional perspective on, 117–119
  program implementation, 119–123
Chemical-heavy farming techniques,
  178–179
City street food, Amritsar, 55*f*
Climate change
  adaptation, 17
  modeling, 14–15
Climatic extremes, 3
Collateral damage, 101, 102*b*
Committee on World Food Security (CFS),
  18–19
Community-based management of acute
  malnutrition (CMAM) model, 10
Community development, 52, 54
Community forestry intervention, 38
Community ownership, 38–40
Community resilience, 27
Conflict-sensitive approach, 93
Conservation Agriculture, 185–186
Consultative Group on International
  Agricultural Research (CGIAR),
  136
Consumers, 168
  producers and, 171

Contrary imaginations, 148–149
Corn cobs, in parched field, 161
Crop improvement, 135–137
Cultural anthropologists, 103
Cultural identity, 25

D
Darwish, Mahmoud, 96*b*
Deforestation, 81, 178–179
Degradation, soil, 178–181
Degraded agro-pastoralists, 2
Department for International Development
    (DFID), 19–20
Deserts, food-getting strategies, 49
Developing countries, economic decision
    making of farmers in, 22–23
Development banks, 117–118
Development intervention
    conflict-/political-break with tradition,
        67–68
    expatriates, 56–58
    food-gathering context, 49–51
    food insecurity, 58–61
    interaction with local administrations, 66
    managing expectations, 68–69
    multiethnic national communities, 66–67
    resilience improvement, 52–56
    triangulation
        Kyrgyzstan, 64–65
        Somalia, 63–64
        Sudan, 63
        surprising attitudes of lightly traveled,
            65–66
        water's edge, 64
        Yemen, 65
        Zambia, 64
Dietary energy intake, 8
Direct Shipping Ore (DSO), 133
Disaster risk reduction (DRR), 17, 104
Do-no-harm failure, Philippines, 77*b*
Drip irrigation, 137–138
Drought-tolerant plants, 161
Dryland agriculture, 15

E
East Africa, 83–84
Eastern Nepal, Upper Barclay, 29–30
Eco/agro-tourism, 155–157

Ecological capital, 74–75
Ecological hoofprint, 73
Economic anthropologists, 21–22
Economic Community of West African
    States (ECOWAS), 94–95
Economic rationality, 22, 24
Economic Recovery in Gaza Strip (ERGS),
    90–91
Economic reproduction, 79
Eritrea, barking dogs in, 134
Ethical consumption, 171
Ethnography, 170, 215
European potatoes, 161–162
Ex-officio members, 113–114
Expatriates, 56–58

F
Farmers, 168
    field schools, 38
    first-generation, 203–204
Farming community, 14
Farm operations, 21
First-generation farmers, 203–204
Fishing community, 43
Flamingo breeding, 149
Fodder sale, Sudan, 44
Fog harvesting, 138
Food, 9
    activism, 170
    availability, 8
    crisis, 21
    politics, 73
    safety, 8, 10
Food and Agriculture Organization (FAO),
    8, 14–15
Food-for-work schemes, 10
Food-gathering strategies, 50–51
Food security, 1, 9, 58–61, 93, 211
    absorptive capacity, 53
    adaptive capacity, 53
    in developing world
        causes of, 9
        cross-cutting issues, 11
        manifestations and measurement, 9
        mitigation of, 10
        prevention of, 10
    participation, 212–213
    social component essential, 211

transformative capacity, 53
World Food Summit, 1996, 1996, 8
Food security resilience, 147
chronic food problems, 14
climate change, 14–15
education to, 91–93
livelihood security, 13
peace dividend
    absence of peace, 94–98
    absence of war, 98–103
policy making and planning, 76–80
    East Africa, 83–84
    Sudan, 84
    Tamil Nadu Integrated Nutrition
        Project, 81–83
    World Food Program, in Gambia,
        Peru and Gaza Strip, 84–91
priorities to improve resilience, 104–107
transitory crisis, 14
UNDP report, 13–14
weak institutional environment, 16
Food security strategies
community ownership, 38–40
cultural adaptation of modern Inuit,
    193–197
fishing community, 43
hunting and gathering, 192–193
Inuit
    cultural factors affecting resilience,
        200–202
    hunter-gatherer communities,
        192–193
private enterprise in refugee camps, 46
room cleaner, Gambia, 46
social and cultural obstacle course
    clearing the hurdle, 28–32
    falling at hurdle, 32–38
social capital, 40–41
social networks, 40–41
success breeds success, 40
Sudan
    desert, tomato growing in, 43–44
    fodder sale in, 44
    goatherd in, 41
    police in, 45
    shoe shiner in, 42
traditional lifestyle *vs.* modern settlement,
    197–200

Uganda
    police on Fort Portal Road, 45–46
    police on Owen Falls Dam, 45
Food systems, 72, 74
France, potato promotion in, 161–163

G
Gambia
    food and nutrition security, 85
    fuelwood, wasteful use, 87*f*
    livelihood security, 85
    Ministry of Basic and Secondary
        Education, 85–86
    multisectoral farm model, 86
    School Feeding Program, 85
    World Food Program, 84–91
Gaza Strip
    economic recovery program, 90–91
    SWOT analysis, 150
    World Food Program, 84–91
Gebrselassie, Haile, 171–172
Genesis, 65
Genetically modified organism (GMO),
    175–176
Glib methodologies, 128
Global agricultural productivity (GAP), 175
Global assessment of human-induced soil
    degradation (GLASOD), 182
Global food security, 76
Global warming, 178–179
Goatherd, Sudan, 41
Grasslands, 49
Greenland, 132*b*, 195*b*
Green Revolution techniques, 10
Groundnut scheme, Tanganyika, 131

H
Harrison, D. W., Dr., 168
Harvest Plus, 171
Healthy life, 8
Human capital, resilience strategy
    country-wide context, 205–207
    first-generation farmers, educated cadre
        of, 203–204
    modern development initiatives,
        204–205
    political and economic context,
        202–203

Hunter-gatherer communities, 49, 79,
    211–212, 216
Hunting, 50–51, 131–132

I

Implementers, community, 147
Indigenous knowledge, 128
Intergovernmental Technical Panel on Soils
    (ITPS), 179
Internally displaced persons (IDPs), 104
International Assessment of Agricultural
    Knowledge, Science and
    Technology for Development
    (IAASTD), 180–183
International nongovernmental
    organization (INGO), 38
International organizations, 117–118
International Security Assistance Force
    (ISAF), 218
Inuit
    cultural adaptation of, 193–197
    cultural factors affecting resilience, 200–202
    hunter-gatherer communities, 192–193
    push factors, 195–196
    socio-economy of, 196–197
    traditional diet, 199b
Inuvialuit, 191
IPCC, 195b
Irish Republican Army (IRA), 103

K

Kyrgyzstan, 37–38, 64–65

L

Lateral thinking approach
    corn cobs in parched field, 161
    flamingo breeding, 149
    Nepal earthquake, 1998, 1998, 163–165
    new idea to a community
        differing perspectives, 159–161
        eco/agro-tourism, 155–157
        rapid-impact technical interventions,
            157–159
    population management, 155
    potato promotion, France, 161–163
    SWOT analysis
        strengths, 150
        weaknesses, 150–155

Liberia, 30–31
    sacred sites, 133–134
Libya, 59–60
Linkage between relief, rehabilitation and
    development (LRRD), 93–94
Livelihoods and Food Security Trust Fund
    (LIFT) program, 104
Local administrations, interaction with, 66
Local community, 52
Local knowledge, in building resilience,
    125, 127
Lord's Resistance Army (LRA), 54

M

Malnutrition, 215–216
Market scenes, Bhutan, 114–115f
Middle East Regional Agricultural Program
    (MERAP), 100b, 143
Mid-term evaluation (MTE), 81–82
Ministry of Basic and Secondary Education
    (MoBSE), 85–86
Ministry of Development and Social
    Inclusion (MDSI), 87–88
Ministry of Social Welfare, Relief and
    Resettlement (MSWRR), 121
Mongolia, 140b
Monocrop agriculture, 51
Morales, Rosario, 97b
Mountains, food-getting strategies, 50
Multiethnic national communities,
    66–67
Multisectoral farm model, 86

N

National Center for Agricultural Research
    and Extension (NCARE), 144
National food security strategies, 16–17
National Resistance Army (NRA),
    98–99
Natural Resource Management (NRM),
    182–183
Nepal, 60
    earthquake, 1998, 1998, 163–165
New Partnership for Africa's Development
    (NEPAD), 80
Nigeria, 60
Noncontentious, 1
Non-Gazan communities, 150

Nunavut, 191
Nutrient deficiency diseases, 9
Nutrient oversufficiency, 9
Nutritional anthropology, 136
Nutritional insecurity, 9
  absorptive capacity, 53
  adaptive capacity, 53
  community resilience, 171
  resilience against
    chronic food problems, 14
    climate change, 14–15
    livelihood security, 13
    transitory crisis, 14
    UNDP report, 13–14
    weak institutional environment, 16
  transformative capacity, 53
Nutritional resilience, 215
Nutrition security, 1, 213
  government-owned social transfers, 10
  Inuit's resilience, 200–201
  traditional lifestyle *vs.* modern settlement,
    197–200
Nutritious food, 8, 10

O

Organisation for Economic Co-operation
    and Development (OECD), 53
Organizational anthropology, 78–79

P

Palestine-Israel Journal (PIJ), 101, 159
Palestine Liberation Organization (PLO), 39
Palestine National Food Security Strategy,
    115, 129
Participatory Rural Appraisal (PRA), 128
Participatory variety selection (PVS), 10, 38
Pastoral agricultural groups, 50
Pastoralism, 49–51
Peace activists, 100–101
Peace dividend
  absence of peace, 94–98
  absence of war, 98–103
  conflict-sensitive approach, 93
People's Democratic Republic of Yemen
    (PDRY), 91–92
Personal security, 125
Peru
  Qali Warma in

National School Nutrition Program,
    87–88
  participation in, 89*b*
World Food Program, 84–91
Petroleum Trust Fund (PTF), 167–168
Planners, community, 147
Ploughing on regardless, 178
Poetic anthropology, 96–97
Police
  Fort Portal Road, Uganda, 45–46
  Owen Falls Dam, Uganda, 45
  in Sudan, 45
Policy making and planning, 76–80
  agricultural systems, 76
  East Africa, 83–84
  global food security, 76
  Sudan, 84
  Tamil Nadu Integrated Nutrition
    Project, 81–83
  World Food Program, in Gambia, Peru
    and Gaza Strip, 84–91
Political hurdles, 1
Population management, 155
Potato promotion, France, 161–163
Poverty, 16–19, 211–212, 217
Poverty reduction strategy (PRS), 89
Productive safety net program (PSNP), 16,
    18, 87–88, 104–105

Q

Qali Warma, Peru
  National School Nutrition Program,
    87–88
  participation in, 89*b*

R

Rapid-impact technical interventions,
    157–159
Rapid Rural Appraisal (RRA), 128
Refugee camps, private enterprise in, 46
Regional groupings, 117–118
Resilience, 1, 3, 53*b*, 211
  defined, 13
  improvement, 52–56
  vulnerable populations, 16
Resilient food systems
  alternative food strategies, 176
  challenges, 173–174

Resilient food systems *(Continued)*
    social cohesion and socio-economic
        integrity, 186–190
    soil fertility, 174
        conventional modern agriculture
            scenario, 177–185
        in perpetuity, 185–186
    soil productivity, 174
Room cleaner, Gambia, 46
Royal Air Force (RAF), 28–29

S
Scaling-up Nutrition (SUN) Movement, 10
Self-evident concepts
    Arctic Inuit, 131–132
    barking dogs, Eritrea, 134
    best of new to old, 142–144
    best of old to new, 138–142
    crop improvement, 135–137
    drip irrigation, 137–138
    feedback on project, Central Asia,
        134–135
    groundnut scheme, Tanganyika, 131
    sacred sites, Liberia, 133–134
Self-help groups, 153
Shoe shiner, Sudan, 42
Silo mentality, 118–119
Single-tactic approach, 74–75
Sinokrot Global Group (SGG), 169
Sinokrot, Mazen, 169
Slash-and-burn techniques, 188
Slum life, Kolkata, 4f
Social and cultural obstacle course
    clearing the hurdle
        Ankole district, Uganda, 28
        colleagues' wellbeing in Afghanistan,
            31
        creating space, 31–32
        meeting the ancestors, 30
        motorbike accident, Liberia, 30–31
        return of king, 28–29
        Upper Barclay, Eastern Nepal, 29–30
    falling at hurdle
        African colonial period, 33–35
        diplomatic incident without trying,
            Kyrgyzstan, 37–38
        sensitivities to being beneficiaries,
            35–37

Tanzanian's sensibility, 33
        Tripoli, Libya, 32
Social anthropologists, 103, 187,
    214–215
Social safety nets, 104–105
Socio-economic conditions, of population,
    60–61
Soil degradation, 178–181
Soil fertility, 174
    conventional modern agriculture
        scenario, 177–185
    in perpetuity, 185–186
Soil organic carbon (SOC), 183, 186b
Soil productivity, 174
Somalia, 63–64
Street food seller, Kolkata, 55f
Stylosanthes, 61–62
Success breeds success, 40, 172
Sudan, 60, 63, 84
    fishing community, 43
    fodder sale in, 44
    goatherd in, 41
    livestock starvation graveyard, 72f
    police in, 45
    rapid-impact technical interventions,
        159
    shoe shiner in, 42
    tomato growing, in desert, 43–44
Sudanese pounds (SDG), 44
Supporting Horn of African Resilience
    (SHARE), 17–18
Sustainable food system, 216–217
SWOT analysis
    Gaza Strip, 150
    strengths, 150
    weaknesses
        at administrative and management
            level, 153–155
        in community, 150–153
Synergies, 111

T
Tamil Nadu Integrated Nutrition Project
    (TINP), 81–83
Tanganyika groundnut scheme, 131
Team work
    advantages of, 112–114
    getting it right as, 115–116

getting it wrong as
  start of Boko Haram, 116–117
  territorial integrity, 116
Tea pickers, in Darjeeling, 4*f*
Terms of reference (ToRs), 90–91, 120
Territorial integrity, 116
TINP. *See* Tamil Nadu Integrated Nutrition
  Project (TINP)
Tripoli, Libya, 32
Trust Fund arrangement, 120–121

U

Uganda, 60, 153
  Ankole district, 28
  police on Fort Portal Road, 45–46
  police on Owen Falls Dam, 45
UK Committee on Climate Change
  (CCC), 184
Undernutrition, 8, 11, 93, 131, 211–212,
  216, 218
United Nations Relief and Works Agency
  (UNRWA), 90
Unsustainable food systems, 72–73
Upper Barclay, Eastern Nepal, 29–30

V

Value Based Voucher (VBB), 90–91
Village animal health workers (VAHWs), 38
Village extension workers (VEWs), 38
Vulnerability analysis, 17

W

Water scarcity, 180–181
Water's edge, Africa, 64
Well irrigation, 63
Windshield survey, 53–54
Women's Development Centre, 47*f*
World Food Program (WFP)
  Economic Recovery in Gaza Strip, 90–91
  Gambia, 85–86
  Qali Warma in Peru, 87–90
World Trade Organization (WTO), 186

Y

Yemen, 59–60, 65

Z

Zambia, 64
Zero Hunger Challenge, 2012, 9

Printed in the United States
By Bookmasters